EXPLORING APOCALYPTICA

EXPLORING
APOCALYPTICA

COMING TO TERMS WITH
ENVIRONMENTAL ALARMISM

EDITED BY

Frank Uekötter

University of Pittsburgh Press

Published by the University of Pittsburgh Press, Pittsburgh, Pa., 15260
Copyright © 2018, University of Pittsburgh Press

Manufactured in the United States of America
Printed on acid-free paper
10 9 8 7 6 5 4 3 2 1

Cataloging-in-Publication data is available from the Library of Congress

Cover art: Photographs by Ratana21/Sutterstock.com and iStock.com/pepifoto
Cover design: Nick Caruso Design

CONTENTS

ACKNOWLEDGMENTS

THIS IS THE FIRST book that grew out of the Birmingham Seminar for Environmental Humanities (BISEMEH). It demonstrates what the environmental humanities should do: seize on an unresolved issue and explore ways to advance debates with the intellectual resources of the humanities. The project benefited from a workshop at Bielefeld University that took place with support from the School of History and Cultures of the University of Birmingham. I wish to thank Hannah Smith for translating the articles by Kevin Niebauer, Patrick Kupper and Elke Seefried, and the two articles by Bernd-Stefan Grewe, and Frances Foley for doing the same with Anna-Katharina Wöbse's article. Two anonymous reviewers provided helpful feedback. And it was a pleasure to work with Sandy Crooms and Alex Wolfe at the University of Pittsburgh Press.

My greatest thanks go to the authors, who took part in an academic endeavor with an elevated risk level. You cannot write about alarmism without stepping out of the ivory tower, which is what the environmental humanities should be all about. It also means to aim for an intellectual middle ground that may not exist yet. With the world being as it is, it is quite possible that vested interests will search this book for ammunition in ongoing conflicts. We hope to convince them that, rather than perpetuating long-standing disputes, it is more rewarding to reflect on why we are stuck with certain lines of reasoning. But every book runs the risk of being misunderstood, and misunderstandings about a book on environmental alarmism can be particularly painful. But, all things considered, it would not be the end of the world.

EXPLORING APOCALYPTICA

INTRODUCTION
THE APOCALYPTIC MOMENT

WRITING ABOUT ENVIRONMENTAL ALARMISM

FRANK UEKÖTTER

ONCE UPON A TIME the apocalypse was a topic for special occasions. It was there for wars and other existential emergencies, for a preacher in need of a sermon that really scared the flock, and for the lunatic fringe. Those days are gone in the new millennium. If aliens were to listen in on one of today's news outlets, they would surely diagnose an infatuation with the end of days. Apocalyptic overtones permeate broadcasts from stock market projections to the latest news from the White House, and every political cause seems to ride on the back of some dramatic horror scenario. The alarmist mode has turned into the default mode of political communication.

Western environmentalism has followed its own trajectory when it comes to alarmism. From the 1970s to the 1990s, environmentalism gained a reputation as a cause that was particularly prone to alarmist rhetoric. To cite just one example, Rick Perlstein argues that since the publication of Rachel Carson's *Silent Spring* in 1962, "Environmentalism had sometimes seemed a sort of transideological apocalypticism."[1] But in the new millennium, two trends have challenged this received wisdom. On the one hand, scientific research has painted an ever more precise picture of environmental hazards. While academic uncertainties cast a pall over discussions when anthropogenic climate change first emerged as a political issue in the 1970s and 1980s, today we can talk about the world's climate with a degree of precision and reliability that has turned denial of climate change into an intellectual embarrassment.[2] On the other hand, apocalyptic rhetoric seems to have lost the political thrust of former years. The 2009 Copenhagen climate summit followed up on a veritable barrage of alarmist rhetoric from the authoritative reports of the UN's Intergovernmental Panel on Climate Change to Roland Emmerich's disaster movie *The Day After Tomorrow*, and yet the event turned into the greatest debacle of global environmental policy.[3]

Apocalyptic environmental rhetoric has drawn a broad range of comments over the years. Julian Simon challenged one of the leading prophets of doom, Paul Ehrlich, to a famous bet whose outcome is a topic of ongoing discussions.[4] Bjørn Lomborg sought to cut through environmental fears with a deep plunge into statistics. A *Common-Sense Guide to Environmentalism* of 1994 attacked mainstream U.S. environmental organizations as members of a "'crisis of the month' club." In 2013, the French writer Pascal Bruckner published a lament, *The Fanaticism of the Apocalypse*, which takes on an ecology that invokes nature solely as "a stick to be used to beat human beings." Others count on the enduring allure of the environmental apocalypse. After unmasking scientists and scientific advisers who downplayed environmental risks in a painstaking empirical critique, Naomi Oreskes and Erik Conway moved on with a book of fiction that chronicles the upcoming collapse of Western civilization between 2073 and 2093.[5] Asked about what readers should take away from the book, one of the authors expressed his hope that readers "think more clearly about the climate of the future."[6] Horror scenarios remain a fixture in the public discourse on climate change, but they increasingly come with a sense of ambiguity. When *New York Magazine* published an article with gloomy warnings about an uninhabitable earth, it sparked unease as well as a nagging feeling that an alternative narrative was nowhere in sight. "Over the past decade, most researchers have trended away from climate doomsdayism," Robinson Meyer noted in the *Atlantic*, but he was unsure what would take its place: "No one knows how to talk about climate change right now."[7]

Controversies typically centered on matters of legitimacy, political clout, and topical focus. First, was alarmism justified in light of the best available evidence? Was speculation about future events a legitimate endeavor for experts, given that the future is uncertain by its very nature, and if so, what were the criteria for legitimate projections? Second, did horror scenarios really galvanize the attention of people and policymakers, or was that an act of wishful thinking? Third, did environmental horror scenarios grow out of a concern for sustainability, or was the environment camouflage for more sinister motives?[8] But as the environmental heydays of the last third of the twentieth century move into perspective, a fourth dimension is emerging that remains vastly underexplored both within academia and among the public at large: what is the legacy of environmental apocalypticism?

The apocalypse is about the here and now by nature. Horror scenarios typically relate to the challenges of the day, and the full drama is bound to unfold in the not-too-distant future. The urgency of the moment usually renders reflections on long-term effects into second-rate affairs, a matter for antiquarians and literary critics who may eventually seize on the matter when the dust has settled. As Frank Kermode has noted in his seminal *The Sense of an Ending*, "Apocalypse can be disconfirmed without being discredited."[9] But in the twenty-first century, it is becoming increasingly obvious that the legacy of past environmental alarms matters for the challenges of the day. The horror scenarios linger, and they shape the ways in which we engage with environmental issues.

In the fable of Aesop, the boy who cries wolf learns a powerful lesson: those who lie will eventually receive punishment. But in the real world, the moral bottom line is far messier: tropes remain in circulation long beyond their prime, and their effects go in many directions. They no longer have the meaning that they used to have, let alone the urgency, and yet people find it difficult to reflect on topics without this legacy. As Patrick Kupper and Elke Seefried show in chapter 3, the Club of Rome's study *Limits to Growth* lingers in debates over resource scarcity, along with a vague notion that it was not all quite so dramatic, and similar statements can be made for the Amazon rainforest, forest death in Germany, and the great upcoming air pollution disaster: tropes barely change, and to the extent that they do, they are just fading from memory rather than being digested and replaced by more sophisticated views. The heat of apocalyptic debates gives way to a strange afterglow, and the fragments of past received wisdoms live on in an undetermined, zombie-like state.

Zombies are inherently destructive creatures, and the same can be said about the legacy of environmental alarmism. As it emerges in the following chapters, their most important legacy is that they constrain the environmental imagination and curtail the range of options. Germans are unlikely to have another public debate about the future of their forests as long as the stir over forest death remains a living memory. The fate of the tropical rainforest struggles to make the evening news, and certainly not for lack of drama. In spite of a plethora of prophecies about peak oil, peak uranium, or peak phosphorus, no vision of upcoming scarcity has achieved the resonance of the 1972 *Limits to Growth*. Anna-Katharina Wöbse (chapter 6) notes that Greenpeace has not achieved another success on a par with the 1995 Brent Spar campaign, and surely not because it did not try.

Of course, this state of affairs has a range of causes. Environmentalists have put much of the blame on neoliberalism, globalization, and public apathy, and the authors of this volume do not dispute that these factors play a role. However, the wisdom of hindsight has not rendered the legacy of environmental alarmism moot: we know that Germany's forests did not perish and that Los Angeles never experienced mass death from an air pollution episode, and yet terms and visions live on in minds and conversations. Unlike other types of undeads, postapocalyptic tropes have remained strangely impervious to the sunlight of enlightened critique.

The reasons for this resilience deserve a more comprehensive discussion, but one factor stands out from the following chapters: environmental horror scenarios imply a dramatic and fateful simplification. All the following stories touch on a significant problem, and yet the exact properties of these problems were fading from view behind the hegemonic cliché. The air pollution disaster trope merged cancer fears, nuclear fallout, and the hugely different events in Donora and London into a diffuse yet terrifying threat of mass death. The *Limits to Growth* left no room for differentiation and did not specify agency. The forest death scare treated all German trees the same, and so did the wood scarcity trope two centuries earlier. Forests were diverse, but the forests under threat were homogeneous. Bringing complexity back in was a challenge, and it did not make much headway beyond expert circles. "We are doomed" is a powerful paradigm. "It's complicated" is not.

The simplistic nature of horror scenarios helped to increase their impact. Simplicity opened the door for related issues to attach themselves to the cause: a precise, academically sound scenario was far less open to associative thinking than an unspecific, blurry threat. As a result, conflicts over horror scenarios were about much more than the core issue. Fear of an upcoming air pollution disaster helped to stimulate environmental protest around 1970, which in turn shaped landmark federal legislation that forms the backbone of U.S. environmental policy to this day. The *Limits to Growth* allowed alternative understandings of economic and social progress to flourish. Forest death marked the breakthrough of green issues in the Federal Republic of Germany, and the Brazilian rainforest played a similar role globally around 1990. The Plachimada conflict, which Bernd-Stefan Grewe chronicles in chapter 8 of this volume, raised awareness of environmental conflicts and environmental understandings in the Global South. These wider implications in turn shaped the course of debates. Retrospective conflicts over the *Limits to Growth*, the forest death trope, or Plachimada were typically proxy wars.

The resilience of alarmist tropes stands in marked contrast to the brevity of their formation. The wood scarcity trope is an exception, as it grew out of the structural conditions of early modern statehood. But for all the case studies between the Second World War and the year 2000, we can observe a phenomenon that one might call the apocalyptic moment: apocalyptic tropes were defined within a remarkably brief period of time. The precise length inevitably depended on the specifics. It was a matter of weeks in the case of the Brent Spar campaign whereas the *Limits to Growth* was not framed as an oil scarcity warning until the 1973 oil crisis shocked the Western world in the year after publication. Narratives can spread and change for a while, but the apocalyptic moment comes to an end after several years at the most, and the defining tropes, or at least some of them, become largely immune to criticism and change. They limp along and continue to bite, leaving people just as hapless as movie actors in the face of a zombie.

As it turns out, matters were not clear-cut at the outset either. A popular trope holds that activists are somehow masterminding environmental alarmism, but that assumption falls flat in the essays of this volume. It does not even make sense in chronological terms. As Bernd-Stefan Grewe shows (chapter 1) in his discussion of the wood shortage scare, environmental horror scenarios were already around at the dawn of modernity. Urbanites in Los Angeles feared asphyxiation long before environmental campaigning became a fine-tuned machine. In fact, in the one chapter where this book traces a campaign organized by a professional, experienced nongovernmental organization— the Brent Spar campaign led by Greenpeace—the campaign went off script in dramatic fashion. The oil rig was intended as a symbol for North Sea pollution in general, but when the general public followed the campaign with growing enthusiasm, the symbol turned into the actual issue, and Greenpeace did not dare to push back.

Of course, horror scenarios are not disembodied tropes beyond agency and interests. But as they emerge in this volume, apocalyptic scenarios typically thrive at the crossroads of several overlapping trends. The wood scarcity trope established itself on the back of the fiscal interests of early modern states, a burgeoning cadre of foresters, the monetarization of the economy, and changing patterns of use. In fact, while early modern statehood was generally on the winning side of the wood scarcity debate, the trope was also used *against* rulers on occasion. The *Limits to Growth* drew on the contemporary fascination with planning and management and the nimbus of the computer. Scientists, journalists, activists, and a sagging economy made forest death a household term in Germany. As Kevin Niebauer shows in chapter 5, the endangered rainforest was a multigenerational project with a range of actors in Brazil and beyond. In short, apocalyptic tropes defy ownership more often than not.

Every apocalypse is immediate to God, but we can identify some recurring patterns in the stories at hand. Experts have played a prominent role in alarmist debates ever since the wood scarcity trope turned forestry into a respectable profession. Journalists were involved in most of the following stories, with some interesting changes in their precise role. They were more of a conveyor belt into the 1970s, an open medium but not an active agent, but the forest death debate of the 1980s saw activist journalists who consciously nurtured terms and tropes. In the case of Brent Spar, the predilections of journalists, and particularly their penchant for dramatic pictures, were a crucial part of campaign planning. The conflict between the Adivasi and Coca-Cola (chapter 8) presents yet another role of the media, as international reporting was crucial for the campaign's success. While news coverage turned a local conflict into a global story, it was remarkably careful not to overburden the story with apocalyptic fears. For all the international attention, the struggle remained rooted in the realities on the ground.

The economy played a crucial role, though scholars from a literary studies background are typically reluctant to acknowledge it. It is striking how the timing of environmental alarms coincides with the great socioeconomic crises of the postwar years. The *Limits to Growth* thrived on the back of the economic malaise of the early 1970s and in turn shaped perceptions of that malaise, as a preoccupation with growth gave way to a preoccupation with limits. The German forest death debate was a reflection of the economic crisis in the wake of the second oil price shock. Brent Spar received an enthusiastic response in Germany because it offered an outlet for the frustrations of a sagging post-reunification economy. Scholars are rightfully wary of the shallows of economic determinism, and yet it may be difficult to explain the cycles of alarmist rhetoric without this context.

We can also observe a notable shift from substantial to merely symbolic themes. The prospect of wood scarcity was a genuinely terrifying prospect in preindustrial economies. Los Angeles became the birthplace of apocalyptic pollution rhetoric in postwar America because the city relied on clean air. The *Limits to Growth* raised a crucial issue when it criticized the obsession of postwar societies with growth. Things look more ambiguous in the forest death debate. Running visions of forestry were mostly framed from a distance, as the real problems of the woodlands took a backseat to the predilections of urbanites who imagined an idyllic sylvan refuge. In the case of the Amazon rainforest, distance was a *conditio sine qua non* for politicization and protest. Brent Spar was never more than a symbol, though those who heeded the boycott against Shell probably thought otherwise in the heat of the campaign. There was an obvious gap between the scenarios that mattered and the scenarios that galvanized the public, and that gap has not seemed to shrink over time.

Of course, this impression is to some extent based on the selectivity of the following set of case studies. At the risk of stating the obvious, this book does not provide a comprehensive assessment of all horror scenarios that environmental history has in store. It does not even look into all major issues that provoked apocalyptic rhetoric, for such a volume would surely need to include an article on anthropogenic climate change, a first-rate generator of popular apocalyptic scenarios since the 1980s. But at the end of the day, this volume suggests that a comprehensive overview may not be such an enticing project after all, as alarmism is not a topic in its own right: it is a feature in numerous environmental discussions and thus best understood in context. At the very least, histories of alarmism should show a familiarity with what was at stake in the interplay between humans and the natural world. It would be a moot point if it were not for so many comments on the environmental apocalypse that show neither familiarity with the issues nor awareness that these issues might matter.

All the following essays explore the environmental challenges at stake, and they come to a clear assessment: none of the scenarios that this book discusses was much ado about nothing. However, it is equally important to note that some of the most popular scenarios were off the mark in small but significant ways. Air pollution did kill, but not in the form of a sudden disaster. Germany's forests were in trouble, though they were not on death row. The oil industry was justly in the environmentalists' spotlight, but disposing of oil rigs was a minor issue. Such a series of narrow misses suggests that there is probably a more fundamental problem at play in the stories at hand. Maybe environmental horror scenarios reflect the inability of modern societies to confront chronic challenges?

But for all the misconceptions, horror scenarios have inspired policies whose retrospective legitimacy is beyond debate. Smog-plagued Los Angeles launched the most aggressive drive against air pollution in postwar America, German power plants were retrofitted with sulfur scrubbers in record speed, and international attention helped Brazilian conservationists. And yet the window of opportunity was surprisingly small: after several months, or a few years at the most, even the most popular horror scenarios were losing their sting. Sometimes environmentalists even found themselves struggling against well-meaning but poorly conceived legislation that was drafted in the heat of environmentalist furor. Frank Uekötter shows (chapter 2) that a 1969 ballot initiative of the Hollywood-based People's Lobby Inc. almost wrecked Californian air pollution control.

Political success relied on hidden requirements. As Frank Uekötter and Kenneth Anders show (chapter 4), it was a closed-door decision of a farsighted official, Peter Menke-Glückert, that pushed coal-fired power plants into the spotlight—otherwise, the environmentalist fury might have flared out in all sorts of directions. Stressing these hidden factors is all the more important since a recent book-length study of the forest death debate does not mention the document that revealed this strategy and instead emphasizes the power of public opinion.[10] A popular myth suggests that it was the raw thrust of outrage and protest that propelled environmental issues onto the political agenda in the 1970s and 1980s and brought about change, and it is high time that environmental historians challenge these fairy tales of the great environmental awakening. Protests did matter, but so did policy brokers behind the scenes, and it takes thorough archival research to identify the latter.

The Cold War is another recurring theme in the following chapters. Environmental historians have explored a range of different ways in which the

Cold War context shaped environmentalism, but when it comes to horror scenarios, one point stands out: the specter of thermonuclear war was a powerful template for apocalyptic environmental fears.[11] But the Cold War eventually came to an end, and it is tempting to speculate whether the environmental apocalypse has lost some of its thrust as a result. The trajectory of the climate change discourse, where apocalyptic tropes are omnipresent and yet strangely powerless, may relate to this change in context. In retrospect, the Cold War years look like the heydays of apocalyptic environmentalism, where tropes and terms thrived with a vigor that the last quarter century found impossible to match. For all the resilience of apocalyptic rhetoric, we may be beyond peak apocalypse.

The slow demise of the environmental apocalypse may also relate to a second trend: the globalization of environmentalism. This volume concludes with two essays on environmental struggles beyond the Western sphere, and these essays provide helpful insights into the limits of Western environmental rhetoric. In chapter 7, Shalini Panjabi traces the multiple emergencies around Dal Lake in Kashmir, while Grewe (chapter 8) dissects a conflict in the southern Indian village of Plachimada, and for all the obvious differences in location and analytical focus, they come down to a joint conclusion: in the Global South, Western-style environmental alarmism does not seem to make much sense.

To be sure, the situation in Plachimada and around Dal Lake was nothing if not alarming. The Kashmir Valley suffered from multiple overlapping emergencies, and Coca-Cola posed an existential threat to livelihoods around its South Indian plant. But the crucial concern in both locales was about reliable information: around Dal Lake, uncertainty reigned even about whether the surface area was actually shrinking. Neither place needs a grand narrative that dwarfs local concerns, and both have very concrete ideas about the way forward: pollution control, adjustments in water management, gradual improvements. People on the ground never sought a short-term campaign with apocalyptic rhetoric, as it was unlikely to achieve much. Places of concern are not necessarily fertile ground for dystopias or grandiose hopes. In the face of multiple emergencies, most people around Dal Lake prefer to live with them as best they can, rather than dream of a life beyond fear.

The experience of the Global South has challenged our perspective on environmental apocalypticism since 1945. Environmental alarmism was neither inevitable nor invariably helpful on the ground. In fact, it may not have been all that popular after all. When Paul Baudrillard met with French environmentalists in 1978, it was Baudrillard who indulged in apocalyptic scenarios, shifting between concerned and humorous moods, while the environmental activists preferred to talk about politics and social change.[12] The Club of Rome was not an environmental organization, let alone a grassroots initiative, and Dennis Meadows's work at the Massachusetts Institute of Technology was about the intricacies of computer-based modeling rather than the counterculture. In the case of the forest death debate, environmental activists did not jump on the bandwagon until it was rolling, as they were typically more concerned about nuclear reactors than coal-fired power plants. And then there was the day-to-day work in pollution control, nature protection, urban renewal, and many other local and regional struggles where information and leverage were the crucial political resources. Apocalyptic rhetoric was no tool for every purpose, and it was just one of many strands that came together in that historic moment that Joachim Radkau has called the age of ecology.[13] And even when apocalypticism emerged as a defining trait, environmentalists were often more on the receiving end: alarmism came to them in the form of academic studies or media reports, and they had to relate to it somehow. In short, blaming environmentalists for an infatuation with the apocalypse has been beside the point more often than not. Paraphrasing Bruno Latour, one might say that environmentalists have never been apocalyptics. They were just part of the game.

Alarmist rhetoric is unlikely to disappear from the environmental discourse anytime soon. In fact, it would be worrisome if it did disappear: in a society that seems to crave its daily dose of apocalypticism, confining the environmental discourse to raw data and academic models would be tantamount to confining it to insignificance. Public debates need popular understanding of complex findings, and popular understanding needs simplification, dramatization, and visualization, and yet those who engage in these debates are invariably standing on a slippery slope. Alarmism is always a matter of degrees, and reflections on how far one should go will be crucial for scientists, activists, and policymakers alike. And as this volume shows, reflections of this kind are certainly nothing new.

But there is one aspect that typically escapes attention: we make these reflections in the shadow of the past. Our engagement with the environmental apocalypse is shaped by a legacy that was framed decades ago. For most environmental issues, the apocalyptic moment was a long time ago, and we will live with the outfall for the foreseeable future. Sometimes that legacy will be a helpful precedent. Sometimes it will be a liability. And sometimes it will be a frame of reference that is neither good nor bad but impossible to exorcise.

Perhaps a future generation of scholars will come to a point where they can decide with the wisdom of hindsight whether, all things considered, environmental alarmism was helpful or not. But peak apocalypse is recent, if it is recent at all, and no such viewpoint is accessible to the authors of this volume. Our goal is more modest: we seek to map a legacy and follow the chain of events all the way back to apocalyptic moments that resonate to this day. We may not come to terms with environmental alarmism anytime soon, but we will never get there if we do not reflect on the path that we have taken.

1

POWER, POLITICS, AND PROTECTING THE FOREST

SCARES ABOUT WOOD SHORTAGES AND DEFORESTATION IN EARLY MODERN GERMAN STATES

BERND-STEFAN GREWE

IN HIS 1836 DESCRIPTION of farming in Westphalia and Rhenish Prussia, the agrarian reformist Johann Nepomuk von Schwerz reported on the worrying state of the forests in the Eifel region:

> One should look on it and weep! A country like the Eifel, where there is no shortage of space, where the soil is, in part, of no use to other forms of agriculture, because it is lacking in dung and fertilizing material, there, on every side, the mountains raise their naked heads, which are covered by no shrubbery, and where no little bird can find a sheltered spot for its nest. This is why the cold north and the bitter northeast winds rage, this is why the rainwater which runs from the peaks is but meager and brings the valleys no relief. Were one to have even so much excess wood that one had to burn it simply for ashes, even this would be a great blessing for cultivation; yet far removed from such abundance, in most places a resident of the Eifel no longer has even the necessary fuel, and he must buy it.
>
> And what then when, in a few years, there is no more wood to buy? We are hurrying toward this sad time with giant strides.[1]

Schwerz was not alone in this drastic depiction of the Eifel. Numerous documents, as well as contemporary images like the landscape paintings of Fritz von Wille, testify that areas that today are thickly wooded often had the appearance of wasteland until the middle of the nineteenth century. Under Prussian rule, many of the barren areas were forested once again, and today's hikers will only be able to imagine the former state of the landscape with great difficulty.[2]

As a fundamental resource of pre- and early industrial economic life, wood was not to be underestimated: wood was not only the main source of energy (and heating) for industry and the general populace. Without wood, food could not be cooked, bread could not be baked, meat could not be smoked, pottery could not be fired, and salt could not be produced. Without the charcoal from wood, iron could not be smelted and forged. Without the by-product potash, textiles could not be washed and dyed, soap could not be made, and glass could not be melted. As a material for making things, wood was likewise indispensable; even specialized woodworkers were often trained in the specific properties of particular types and species of wood. For example, not every type of wood was suited to furniture building or wagon making. Some woods like oak withstood ground humidity so they resisted rot better than wood from other species that could not be used to make posts. Most implements in the home and in industry were also made of wood. Finally, as a building material wood is still in use today. Not only did half-timbered buildings and many rural farm buildings consist largely of wood, a great deal of wood was also used even in stone buildings: ceiling beams, windows and doors, floorboards and parquet flooring, stairs, and roof timbering were made of wood. In brief: wood was indispensable. If a wood shortage became acute, every section of society was directly affected—albeit to a varying extent. Therefore, for people in the preindustrial era an impending wood shortage was a genuinely terrifying prospect.

The dwindling of available wood resources, which Schwerz was not the only one to predict, represented a grave danger for an early industrial upland region like the Eifel. Since large quantities of wood could only be transported relatively affordably by water, the import of wood from other areas was difficult. Unlike the British Isles or the Netherlands, in this landlocked, hilly, and mountainous region, waterways were more useful for exports than for bringing in supplies. If wood was transported over land, it was unaffordable for most of those who used it. On these grounds alone, the region's population had a keen interest in the preservation of local wood reserves.[3]

In the opinion of the Prussian historian Heinrich von Treitschke (1834–96), the lamentable state of forests in the Rhineland was entirely the fault of foreigners—that is to say of the French:

> A nigh hopeless task arose for the new government from that dreadful devastation of the forest, for which the forest-loving Teutons were least able to forgive the Latins, among all their sins. The Bergisch farmer clenched his fist if one spoke to him of the old pride of the country, the King's Forest, and the Frankish Forest. Of all the hundred-year-old oak and beech trees, not one remained; and what this destruction of the forest meant for the climate and the cultivation of the blustery heights of the Hunsrück and the Eifel could now only be understood with horror, when suddenly after a storm the mountain torrents plunged into the Moselle Valley and in a few moments washed away the fertile soil, which the poor vintner had labored for months to carry up the shale crags. This monstrous devastation was never to be quite healed.[4]

Treitschke did not concern himself with the actual causes of desertification; for him the point was to find another example of poor management of the regions that had been temporarily ruled by the French. However, even if you engage with the question of deforestation in the preindustrial era in a less narrow-mindedly nationalistic fashion, you have to acknowledge that as far back as the late Middle Ages people complained in countless sources about the devastation of forests and the lack of wood.

It was not only the ruling princes and their foresters who were concerned about the forests; consumers of wood, from bakers to the owners of iron works, saw their existence threatened by the wood shortage. From Upper Bavaria to East Prussia, from Baden to Bergisches Land, in free imperial cities and in newly flourishing trade centers, complaints about the lack of wood rang out. In the early modern period, wood shortages and ravaged forests were subjects of constant complaint almost everywhere in the German-speaking world. Around the end of the eighteenth century, competitions with hefty prize money were held to find out how the wood shortage could best be remedied. A torrent of texts about the destruction of forests, the lack of wood, and ways to use wood more sparingly flooded the German book market.

Since Tacitus, Germany has been considered a richly forested country, and today around 30 percent of Germany's land area is still covered by woods. Thus, it is clear that people in the past were successful in halting the feared destruction of the forest, or in countering its effects. The forests have recovered again, and cover a greater area today than they did on the eve of industrialization. Does this mean the "wood shortage" was a false alarm? Or did the alarm perhaps trigger the right reactions so that, through a change in behavior, the onset of a real wood shortage could be prevented just in time? How should we judge the general lament about the lack of wood?[5]

This chapter seeks to address these questions. The discussion of the "wood shortage" arose particularly frequently in a political context, as the rulers of territories used this argument to try to assert their control over the forest against others who used it. Complaints about wood shortages and the destruction of forests were an instrument to impose political control over the woodlands.[6] However, wood users also took advantage of the argument and complained about the lack of wood. In these cases, the grievances can mostly be understood as an appeal for intervention to the ruling power, whose help they wanted in securing their own supply of wood against competing interests. The "wood shortage" was intensively debated in the public sphere, but was it a case of an anticipated wood shortage, or one that was already being felt? The main argument of this chapter is that although wood shortages were a real phenomenon, they were limited in their social and temporal dimensions. A general wood shortage did not occur but the claim of it was an anticipated crisis.

THE "WOOD SHORTAGE" ARGUMENT AS A LEGITIMIZING FORMULA

The Württemberg Forest Ordinance of 1540 already proposed "to hinder the decline and lack of wood, so certain and thus clear to the eye, and make the forest grow, rise and multiply."[7] In a later forest ordinance from the same territory (1552), the local ruler exercised some self-criticism. His government had been "far too mild" ("vil zu milt"), "so that the woods and forest have fallen so far into troublesome and harmful decline, which, if not countered well and in time, the terrible and damaging lack and dwindling of wood, and other faults and disorder, growing longer and deeper every day, will be talked of by us, our country and people, vassals and associated people, and for posterity."[8] Until the end of the eighteenth century, concern about the preservation of forests and the supply of wood for future generations became the standard reason for the issuance of forest ordinances.

During the Peasant's War (1524–25), a revolt of about 300,000 peasants, the rebels' "Twelve Articles" demanded the return of all the forests that had been acquired illegitimately by the ruling princes.[9] But the failure of this biggest revolution in German-speaking Europe enabled the rulers to significantly expand their control of the forests. Over the course of the seventeenth century, almost every German territory issued a "forest ordinance," often with similar wording. At first the prohibitions and laws that were contained in these ordinances extended only to sovereign's forests, but from the eighteenth century onward they increasingly included intervention in the property of other owners.[10] The older forest ordinances mostly issued a list of bans, in particular on deforestation and clearing. The grazing of goats, which was particularly damaging for young trees, was also often forbidden.

Typical patterns of reasoning appeared in the different forest ordinances, which recurred again and again over the following two centuries. For example, in the Forest Ordinance of the Monastery of St. Blasien (1766), it states in the preamble: "The reason for composing this forest and woods ordinance is that the lack of wood, which will soon be apparent in all places, penetrates little by little into the very depths of the Black Forest."[11] Among the several hundred forest ordinances that were published and implemented before the end of the Ancien Régime, most used this kind of legitimizing formula. The traditional forms of using the forest were now increasingly limited, certain forms were banned, occasionally the operations of individual industries with particularly high wood consumption, such as glassmaking, were prohibited. This was often opposed by industries or peasants who used the forest. For this reason alone, the destruction of forests and the lack of wood were frequently directly attributed to the idea that permits for use (which often established usufructs) were misused or that previous ordinances had not been properly observed. At the center of these regulations stood wood production, which was not supposed to be increased through targeted economic measures, but instead protected through bans on competing uses.

How should we judge this form of legitimization, which was used so frequently? Is it credible if, for example, in a 1666 Forest Ordinance of the Electorate of Cologne, the principal reason given for the decree was the wood shortage? So soon after the Thirty Years War, with the dramatic fall in population and the collapse of iron-making, such a complaint seems highly unrealistic. Furthermore, the text matched an earlier ordinance from 1590 almost word for word.[12] This significantly strengthens our impression that this was a constantly repeated formula, and it is thus reasonable to have doubts about the extent to which these preamble texts corresponded to reality.

How these forest ordinances should be assessed is disputed in forest history. Even their frequent repetitions can be interpreted in different ways. They can be seen as "laws that are not enforced" (something that Jürgen Schlumbohn saw as a "structural feature of the early modern state") and that were therefore often ignored.[13] Or they can be seen as a code of conduct decreed from above, which was supposed to be drilled into the memory of the subjects through frequent repetition, and which generally corresponded to the facts and was mostly observed.

Whereas traditional forest historiography sees forest laws as among the most important types of sources,[14] the validity of what they tell us about the state of the forest has been viewed much more critically by other historians. These researchers point out that until the reform era of the early nineteenth century, local rulers did not always have the right tools to enforce regulations where they applied in the forests. Joachim Radkau, Ingrid Schäfer, and Joachim Allmann emphasize the nature of the forest laws as instruments of power, which were intended not so much to protect the forest as to impose and consolidate sovereignty. The ordinances sought to establish control and to make the protection of resources dependent on the ruling power. These scholars have fundamental doubts as to whether complaints about "wood shortages" and "devastated forests" really have anything to do with inadequacies of supply for the general population or for industries. They argue that, for the authorities, forest ordinances were not about the forests but about the preservation and implementation of "order."[15] If this critical approach seems too narrowly concentrated on the interests of the ruling powers, it is nevertheless doubtless that the forest ordinances should be seen in the context of early modern era "police ordinances" and "police legislation," which intended a stronger surveillance of public life and administration. For they reflect not only a general desire for regulation but, more important, an increasing trend toward fixing the actions of the rulers into formal norms. Forest ordinances did not just alter people's dealings with the forest, they also picked up on norms that had already been established by custom and sanctioned them.[16]

In the eighteenth century and in the forest law of the early nineteenth century, alongside thoughts of official "order," a defense against "indiscriminate"
behavior also appeared. On these grounds the state forestry commission justified its interventions into communal forest management, the forced divisions between cooperative, community, and common-land forests and the
supervision of privately owned woods. Individual interests should come second to the concern for the greater good. A typical example of this pattern of
argument is the justification offered by the Prussian state forest master, Georg
Ludwig Hartig, in 1816 for a restriction on communal forestry by the state:
"Wherever this indiscriminate cultivation takes place, and there is likewise
no lack of wood sales, communal and private woodland has either completely
disappeared or fallen into the most sorry circumstances . . . large swathes of
land which had an overabundance of wood 100 years ago are now so poor in
it that one uses heather, straw, and even cow-dung to protect oneself against
the cold in winter."[17] According to this pattern of thinking, only the territorial sovereign or the state was in a position to override individual interests
and safeguard the greater good—in other words, only they could prevent
indiscriminate usage and the accompanying destruction of the forests and
impending shortage of wood.

The critical perspective on the attempts by sovereigns to establish norms
has certainly brought to light some extremely self-serving interests on the part
of the territorial rulers. Even if they were not able to impose their order on the
forest comprehensively, the fear of the sometimes hefty penalties for violating
the ordinances ensured a certain recognition. The many violations were an
important source of money for the rulers; many foresters virtually lived off
their share of the fines that were imposed. From the middle of the eighteenth
century onward, in most German territories the sovereigns' financial interests
had the upper hand over their interest in hunting in the forest. Initial analyses
of the Landgraviate of Hessen, the Bishopric of Würzburg and the Electorate
of Trier show the importance of income from the forests to the overall revenue
of the sovereigns.[18] Likewise, in the post-Napoleonic era, income from the forests was indispensable for most German states; in the Kingdom of Bavaria, for
example, it accounted for 8.3–16.0 percent of all annual state income.[19]

These financial interests in woodland remained a strong motive for state
activity in the forest sector until well into the twentieth century. Sovereignty
and state played a double role here: on the one hand, they set the legal conditions that applied to all forest owners and made sure they were followed; on

the other hand, they had strong economic interests of their own in the forest, because they were in a relationship of competition with the other owners.

But the argument that there was a wood shortage was not only used by territorial rulers to legitimize their claims to regulate the forests. Soon other forest users, who were affected by attempts at regulation, employed the same arguments about "wood shortages," "the destruction of forests," and "indiscriminate" behavior (on the part of the authorities), and used them in certain conflicts against the territorial rulers to pursue their own interests.

In the Electorate of Trier, it was initially the regional estates (*Landstände*), in particular the representatives of the cities and the monasteries, who used the argument about wood shortages in their conflict with the elector's hunting and forestry commissions. The forestry commission was accused of having created a vicious circle by imposing excessively harsh penalties on rule-breakers in the forest, and of acting purely out of financial interest. Through the proliferation of fines: "The misery of the populace and the general shortage of wood grow in a perpetual progression."[20] The Electorate of Trier's estates lamented that the forests, by that point under the control of the forestry commission, were "now in the worst possible state."[21] In fact, this conflict was about who should have control over the felling of timber in nonsovereign woodlands. Both the forestry commission and the estates used the "wood shortage" argument here, and blamed the other side for the problem. The same pattern of argument was used by communities in conflict with the authorities over the introduction of *Schlagwaldwirtschaft*, a forestry method whereby trees are grown and felled in certain predesignated sections of woodland. It has rightly been pointed out that the goal of reducing the shortage was open to fierce debate, and the argument that there was a lack of wood therefore offered a good basis with which to justify one's own political concerns about the forest. Of particular significance is the line of defense that the forestry commission used to fend off various accusations. To underline the effectiveness of its work, it repeatedly emphasized that the woodland it controlled was in the very best condition. But this in no way hindered the forestry commission from referring in the same article to a generally imminent wood shortage. The local communities also used the wood-shortage argument in a similarly strategic way; to defend against interference by the authorities, they also pointed out the good conditions of their own forests whenever the occasion arose.[22] In the Duchy of Westphalia, it was even the estates themselves that took the initiative on a new ordinance on the use of the *Markwaldungen* (forests under common ownership of several villages). Their stated justification for this action was the "wood shortage which is already present in many places."[23]

In addition to the estates, cities, and communities, individual subjects also used the wood-shortage argument to justify their emigration to other countries. The migrants frequently explained their decision with reference to "rising wood prices" and "shortage." A councillor from Württemberg reported on the causes of migration, summarizing the situation and making reference to lower-ranking departments: "The high price of wood and lack thereof comes of the most thrifty taxes of the forestry commissions, the rise in wood prices, and the limits on wood-cutting days, as well as the numerous prohibitions on the forest and also the protection of common game. Notably the wood shortage is caused not by poor people collecting wood, or by their livestock and pasture . . . but by the quantity of wild game, which does not allow any young wood to grow."[24] In another example, Johannes Kleinefeller, a prosperous sixty-year-old farmer, answered a question from the District Magistrate of Hanau: "What motivates his proposed departure, and thus what causes can he actually state?" by saying: first, because taxes now have to be paid in coins, second, "because there are too many restrictions on logging in the forests," and third because of the restrictions on using water to irrigate fields.[25]

It is by no means certain whether the reasons stated here were actual grounds for migration. Although from the twentieth century onward, the politics of migration in most states has tended to focus on limiting immigration, most principalities of the Ancien Régime were instead interested in having the largest possible population and in preventing emigration. For this reason, during these hearings, those who wanted to emigrate would predominantly use arguments like these, which would be accepted in the face of the official requirement to stay. In this particular context, a specific discourse of justification often developed in which arguments, such as that of a wood shortage that most sovereign rulers explicitly referred to in their forest policies, could hardly be denied.

In summary, it is clear that particularly in political disagreements and when the interests of the ruler and his subjects diverged, the argument that there was a shortage of wood often cropped up. This argument was employed by all parties in the pursuit of different goals. Essentially, a predominantly political question was being debated. The dispute was over who should be able to make decisions about the legitimacy of using the forest.[26]

COMPLAINTS OF WOOD SHORTAGE AS
AN APPEAL FOR SOVEREIGN INTERVENTION

In addition to these complaints of wood shortages, which should often be examined in connection with questions of legitimization, such complaints also appear in the form of a second, likewise extremely common, model. This model was characterized by the way it was directly and openly motivated by economic interests—unlike the financial interests of the ruling lords, for example, which were not publicly acknowledged, even if they were very evident. Wood users, in particular the people who operated iron- and glassworks, saw their wood supply threatened, or already did not receive enough wood for their needs.

Thus, for example, in 1711 the steel hammersmiths threatened the Lordship of Schmalkalden with revenue losses since, with "ruin before their eyes," they would have to halt their production in winter and spring "because of the lack of fir-charcoal that has very much come to pass." They asked to be assigned a larger quantity of fir wood to be turned into charcoal. Although the forest administration wanted to refuse this higher allocation of wood, a royal legislative degree from the following year granted the petition so that the charcoal makers would "not be made to go hungry." But just four years later the smiths complained once again with the same arguments about a shortage of wood, and the negotiations entered a new stage.[27]

In many territories, smelting works, with their high consumption of charcoal, were guaranteed particular terms of purchase and comparatively cheap wood prices through special contracts. When they nevertheless complained about a lack of charcoal, it was generally because they wanted to improve their terms of purchase or defend them against other demands. When the price of wood, and thus also the cost of making charcoal, rose sharply in the eighteenth century, complaints mounted about an imminent shutdown of operations due to wood shortages. When it came to wood prices, the sovereign owners of the forests often complied only to a limited extent, because this would cause them a direct loss of income. In any case, there were other ways in which the complainants could be helped.

A typical reaction, which was often the result of the commercial nature of respective governments' economic thought, was to ban the export of wood and charcoal to other neighboring territories. In Electorate of Cologne-controlled Sauerland, the first charcoal export ban was decreed as early as 1679, and was repeated at regular intervals. Economic links between the Duchy of Westphalia (Electorate of Cologne) and the principality of Nassau-Siegen were badly damaged by this, in particular the trade in crude iron and charcoal, which until then had flourished. For more than a century, these measures were the subject of conflict within Westphalia as well, and here again the argument that there was a wood shortage played a major role. Because they feared interruptions in supply, towns and local smelters were opposed to an end to the export ban, whereas the knights (who themselves more often owned woodland) were very strongly in favor.[28]

In Lippe-Detmold, Nassau-Dillenburg, and very probably in other territories too, the government decided, when it came to the supply of firewood, to privilege certain industries that suited its own fiscal interests over others. In this context, Ingrid Schäfer has spoken of a Late Absolutist "hierarchy of industries" (source term: "Holzmenagirung" [wood provision management]), according to the principle found in a contemporary source that "less important forms of commerce, as well as disadvantageous branches of trade and factories, must give way to the main form of commerce." To protect Lippe's linen industry, for example, the operation of glassworks with which they competed for potash was restricted. Meanwhile, in Nassau-Dillenburg the sovereign and privately owned iron industry was regarded as "the main form of commerce." Private industrial operations that needed wood, for example, brick baking and pottery making, were directed to use alternative fuels in order to save wood for iron smelting. Once again, the familiar argument about shortages was invoked to justify these restrictions on the use of wood or the export bans on wood and charcoal.[29]

Even if ironworks had to halt production because of an alleged "wood shortage," it was often the case that, under a new owner, the same smelting works started production again within a few years; for example, the Hallberg smelting furnace in Nassau-Saarbrücken, which closed in 1773, opened again in 1780.[30]

Observations of this sort led some contemporaries to develop a deeply skeptical attitude toward these complaints about wood shortages: "The already privileged hammer-mill owners say that a wood shortage is at hand, and the supplicant says that there is an abundance of wood, if need be he'll produce proof of it. But as soon as he has obtained what he asked for, he joins in with the general dirge about wood shortages, singing along in perfect harmony."[31] This article from the *Westfälischer Anzeiger* (Westphalian Gazette) shows clearly how wood users who were privileged by the governments of many territories used the argument of a "wood shortage" to keep unwanted competitors away from this resource. At the same time it is clear why, conversely, it could be just as important to deny a shortage that others said was there. Complaints of this kind only made sense in the context of a wood market that was regulated and controlled by a sovereign power. In a free market, such complaints would have been counterproductive because they would have had a negative effect on consumers in the form of rising prices.[32]

Thus, when it comes to economically motivated complaints about wood shortages, we can see that in this context scarcity was likewise used as an argument to gain advantages (in the form of cheap raw materials) through the intervention of the sovereign power. The groups making these complaints varied, as the argument allowed them both to demand special terms of purchase and to shut out competitors. It was therefore also used by subjects, for example, to defend themselves against the arrival of new, large-scale consumers. Although territorial lords referred, as a rule, to a looming wood shortage when they issued their own forest ordinances, at the same time they seldom hesitated to approve new, wood-guzzling smelting works and to endow them with privileges.

POPULAR WRITING ABOUT THE WOOD SHORTAGE

Writings about the alarm over wood shortages, which appeared most notably from the middle of the eighteenth century onward, form their own important category within the scope of this topic. They do not always refer to the "lack of wood" or the "wood shortage" explicitly in the title, but nevertheless countless publications were primarily concerned with the problem of wood supplies in the light of increasing consumption as a result of economic development and rising population. We can divide the proposed solutions into two groups. On the one hand, a string of authors hoped to be able to reduce wood usage (demand) through targeted technical measures or changes in habits. A second solution was favored above all by a newly evolving professional group—the foresters—who wanted to increase wood production (supply).

The first group of writings comprised first and foremost "fuel-saving oven-" and "wood- saving literature." This contained suggestions about how, through the use of more efficient technology, less wood than before could be used to produce the same effect. The debate over the most effective technology encompassed the most varied fields. In the charcoal-making process, for example, significantly less wood could be used by putting logs into the charcoal kiln vertically, instead of piling them up horizontally.[33]

Most often, the question was asked as to how hearths should be set up in people's homes in order to use the smallest possible amount of wood. The Prussian Academy of Sciences in Berlin therefore announced a competition in 1765 "on the building of wood-saving ovens."[34] Also in Berlin, a "Society for the Art of Saving Wood" was even founded in 1784. People tinkered with the designs for ovens equipped with the optimal kind of chimney and a controlled combustion system, which held warm air in the oven for a long time and emitted heat into the room over well-conducting walls. In comparison to an open hearth, the fuel savings must have been considerable, but most of these designs did not catch on, as they were largely too expensive or did not function properly. Thus, a forester from Amorbach scoffed:

> *Even if in all the forests*
> *Wood is very scarce*
> *You can heat your room with bales of books*
> *On how to use wood less*

> *Mag in den Forsten allen*
> *das Holz auch selten sein*
> *man heize mit den Ballen*
> *Holzsparungsschriften ein*[35]

The governments of the various German states were certainly happy to support wood-saving ovens as a solution, because this held out the possibility of being able to pass off the issue of lowering wood consumption onto private households, while simultaneously protecting large-scale industries that were more economically worthwhile. However, since the price of raw materials normally made up more than half the cost of production, people on the side of the smelting works were also always searching for more effective combustion methods. Still, for German iron production, a decisive technological breakthrough did not come until the 1830s with the introduction of the puddling process.[36]

The journalistic debate about the wood shortage was strongly bolstered by a newly emerging profession: that of the foresters.[37] Without a doubt, they made up the group that should have been most familiar with the state of the woods. The foresters used the argument that there was a wood shortage to justify to their respective governments why it was necessary to introduce certain measures to foster the growth of the "ravaged" and "devastated" forest. Often, although by no means always, these suggestions were then implemented both in the law of the forest and in the forestry work that took place there. At the center of the catalog of forestry measures stood a boost in wood production—that is, an increase in supply.

These measures followed two main trends. The first limited agricultural uses of the forest—such as fattening pigs, grazing large livestock such as cattle or horses, growing crops in the woods, raking and collecting leaves and moss to use as animal bedding, cutting and picking grass—and making these activities secondary to wood production.[38] Some uses that were not protected by legal title or were particularly damaging to wood (for example, grazing goats) could simply be banned by law, others were limited or confined to certain areas of the forest. In this way, the growth of young plants was protected from the sickle and from the mouths of hungry herds of animals.

The second trend called for the state of the forest to be improved through active measures. It was argued that new trees should be grown with targeted sowing and planting in forest dells, clearings, and areas of wasteland, and in this way total wood production should be increased. The focus here was on a more effective use of the available land. The most important concept that the foresters produced in the context (coming up with it as far back as the eighteenth century, before it was later converted into a more systematic form by Georg Ludwig Hartig) was forest sustainability. As an economic principle, this idea had already appeared in 1713 in the *Sylvicultura oeconomica* by Hans Carl von Carlowitz. The notion that a generation should not cut more wood than is regrowing in the same period in order to ensure wood supplies for the next generation is doubtless older than this. In some regions with ironworks or mines it has been shown that this form of using the forest was practiced as far back as the late Middle Ages. For instance, this is also true of early conifer plantations, for example, in the Nuremberg Imperial Forest or in the Frankfurt Municipal Forest. The novel aspect of the concept of forest sustainability as it was formulated in the eighteenth century was that it was no longer just focused on small areas, but was supposed to be applied as a general rule to all the woodland in a territory.

It is easiest to monitor and control the sustainable use of wood in a forest if the trees are not extracted individually from across the whole woodland area (as is the case in a "plenter forest").[39] Instead, if logging is concentrated on an area specially set aside for that purpose, overuse (as well as embezzlement by forestry staff) is easier to detect. The logical consequence of this was the introduction of the method known as "Schlagwaldwirtschaft," whereby every year a certain section of the forest was allocated for woodcutting. In order to do this, the forest that was being cultivated was divided into "Schläge" or "felling areas." The amount of logging was supposed to be determined by the planned growth period for the forest, which later was assigned the technical term "Umtriebszeit" (rotation period). If there was a planned growth period of seventy years, then every year one-seventieth of the wooded area (or alternatively one-seventieth of total wood reserves) could be felled, while the remaining sixty-nine areas were left to grow, with a yearlong interval between each one. At the turn of the nineteenth century, getting the highest possible yield of wood from a predetermined area became forestry's main priority.[40]

Foresters had an interest in diagnosing a wood shortage because they thought of their own profession as the suitable remedy. In German forest history, a subject that is very strongly oriented toward its own tradition, the "wood shortage" has therefore become a truism that for a long time has rarely been scrutinized. Indeed, it even often constitutes the founding myth of the subject: "The wood shortage gave birth to forestry," states a standard work on forest history from the 1990s.[41]

Other groups that contributed to the alarm over wood shortages in popular writing were, first and foremost, officials and educated members of the bourgeoisie. Like the foresters, they were sometimes closely connected to the ruling power. Many of the authors who took part in the various competitions held by academies to look into the issue of wood shortages or of how to save wood had also received training in governmental accounting. Therefore, it comes as no surprise that the solutions they proposed were often along similar lines.

In this discourse about devastated forests and a lack of wood it does however stand out that additional uses of the forest, in particular by agriculture, were treated as being responsible for the destruction of woodland. On a linguistic level, these agrarian uses of woodland were treated as less valuable. From the eighteenth century onward, they were referred to more and more frequently as "secondary uses" and by the beginning of the nineteenth century this notion was generally established. The agricultural functions of the forest were also overlooked or diminished in other ways in the debate about forest resources. For example, writings that lamented the devastation of forests only mentioned a shortage of animal feed or manure in exceptional cases—most only talked about the shortage of wood. This also indicates who was *not* taking part in this written debate: the general population practicing agriculture. The importance of the forest in particular to those with small landholdings, who used it to supplement their grazing areas, for gathering animal feed for the winter, and for animal bedding, has been emphasized by recent historical-geographical forest research.[42]

In this context, it is no surprise that the historiographical debate about the wood shortage also concentrated almost exclusively on the "main usage" of timber. Today, the boundary between agriculture and forestry is visible in the landscape. The areas that are used for pasture, for cultivating crops, and for growing trees are separated nicely and neatly, often by a straight line. However, until the last third of the nineteenth century, in many regions farming was only possible if the forest was integrated into the agricultural process as an area of reserve space. Shortages of animal feed were by no means exceptional. A forest superintendent wrote to his superiors in 1830: "The lack of animal food is currently so severe in the countryside, and especially in the *Hochwald* communities, that those who are suffering the most can no longer be prevented from procuring the feed that is desperately needed for the preservation of their livestock through illegal grazing in the young felling areas of the royal forest."[43] Anyone looking through the forest records for further evidence will encounter complaints of this sort very frequently. However, such grievances were marginalized in popular writing about the lack of wood. They were only the subject of extensive discussion in the context of agrarian writing related to the necessary introduction of better crop rotation and more effective production of fertilizer.

In order not to automatically accept the increasing exclusion of the rural population from the forest through one's approach, in my opinion historical research should in future not only examine the wood shortage, but the lack of forest resources from a broader perspective.[44]

THE WOOD SHORTAGE:
BETWEEN RHETORIC AND REALITY

In 1839, the office of the governor of the Prussian Rhine Province sent out an article by an anonymous author with the heading "Preservation of the Forests and Wood Shortages" to its subordinate agencies: "Even though so much has been written and taught about the value of the forests and their care, about their protection and their rejuvenation, the extent of the forest has diminished in the same proportions that the extent of writings being circulated about forestry has increased. We see in particular in those regions that are characterized by agricultural advances, by the rapid flourishing of manufacturing and factories, or by a rising population, that the forests even disappear from land which does not allow for any other uses."[45] In this and other similar articles, the forests in the Prussian Rhine Province (Bergisches Land, Eifel and Hunsrück) were still described as being in a perilous state at the end of the 1830s. In the same year, the governor's office accepted the suggestion of a newspaper to form Forest Protection Societies in all the various administrative districts. To relieve the wood crisis, these societies were supposed to contribute to preventing the infringement of forest laws, to teaching the population how to save wood, and to laying out communal plantations. It is evident how little the Rhine Province's administrative staff thought these kind of societies and measures were needed, based on the fact that it was only after repeated inquiries from the governor's office that they answered and roundly rejected the proposal. In the Rhine-Ruhr region, coal was already supposedly replacing wood as domestic fuel. Likewise, in Koblenz and Trier, people did not believe that wood was being wasted or that there was an insufficient supply of it, since wood prices were said to be very high (Koblenz) and there were "significant quantities remaining for export as well as for the satisfaction of the needs of the industrial establishment" (Trier).[46] As long as regional ironworks could be supplied with charcoal, the authorities responsible for the forests in Trier and Koblenz did not think there was any wood shortage.

However, crime statistics for this period paint quite a different picture (see table 1.1). In no other Prussian province were contraventions against the forest laws as common as in the Rhine Province, and they were even higher in the Bavarian Palatinate and in the north of Baden.[47]

TABLE 1.1 NUMBERS OF WOOD THEFTS IN
PRUSSIAN RHINELAND, 1844–1850

YEAR	1844	1845	1846	1847	1848	1849	1850
Wood	50.650	59.098	54.745	67.029	45.747	58.843	54.560

It was obviously very difficult for many people in the region during this period to provide themselves with enough wood by legal means.[48] It is clear that they suffered from a lack of essential firewood. High wood prices were often an insurmountable obstacle to less affluent social groups trying to provide themselves legally with fuel. But even low wood prices would have been too much for many households that were almost or completely destitute. In this respect, contraventions of the forest laws and wood thefts indicate a social dimension to the wood shortage; not everyone was affected by it to the same extent. While some sections of society had access to enough resources to meet their needs even in times when not much wood was available, others were cut off (even if this was not intentional) from a legal supply. For this reason, the idea of a wood shortage should be considered in a socially differentiated manner; many people could suffer from a lack of wood, even in times when the supply of wood was ample.[49]

Whether there was ever a general and widespread lack of wood, or whether this was only the case in a few regions, is a subject of great disagreement in academic circles. Certainly, the argument that there was a wood shortage must have had a certain amount of plausibility for contemporaries, otherwise it would not have been used in political and economic disputes.[50]

The many assertions of wood shortages and the references in sources to devastated forests have led many historical accounts to equate complaints about a shortage of wood with the real existence of this phenomenon. For example, the economist Werner Sombart, who was world famous at the beginning of the twentieth century, reaches the conclusion in his magnum opus *Modern Capitalism* that around the end of the eighteenth century European industry had been plunged into crisis because of the shortage of wood. Because of the central role played in economic life by wood, he even saw a symbol of "the impending end of capitalism" in the wood shortage.[51] Countless historians have backed up Sombart's view (albeit in a milder forms) and have likewise put forward the thesis that there was a general shortage of wood as well as an energy crisis.[52]

Since the 1980s, a series of researchers have called the truth of this notion into question by highlighting the interests of the sovereign powers in issuing forest ordinances, and by underlining the contradictory nature of the different sources and the instrumentalization of the complaints by different conflicting parties. They therefore call for a more careful reading of the sources relating to wood shortages.[53] As a result of this justified criticism of earlier source interpretation, recent scholarship has reached the conclusion that we should develop a more differentiated definition of what a wood shortage means. These historians call for a clear separation between the discursive phenomenon of a wood shortage and the real phenomenon, since discussion of a lack of wood is by no means necessarily the result of a true shortage. Furthermore, one must clearly differentiate when, in which regions, to what geographical extent, and for how long such a crisis existed, as well as which social groups and industries were affected by it and which particular kinds of wood were no longer available. But despite these shared aims, recent studies have produced quite different results. In his study on the Eifel and Hunsrück regions in the eighteenth century, Christoph Ernst used careful critique of his sources and new methodology to restate the thesis that there was an immediately imminent shortage of charcoal around 1790. He supports this finding predominantly with forest planning data on planting density and on the supply of wood, and he emphasizes the social nature of the wood shortage.[54]

The social dimension of the shortage of forest resources is likewise emphasized in a work on the Bavarian Palatinate in the nineteenth century. In this region, infringements of the forest laws occurred four times more frequently than in the neighboring Rhine Province. But the wood shortage felt by many in the Palatinate was not due to a lack of wood reserves in the forests or a lack of transport infrastructure, but rather was the result of a wood harvest that until the end of the 1830s was even lower than what would have been possible under a strict interpretation of the principle of sustainability. A reason for this was the difficulties that the forestry commission had in calculating the yields from the still relatively heterogeneous areas of forest in which many different types of trees of different ages were grown together. It was only after continual interventions brought the ecological makeup of the forest more closely into line with the categories used for forest planning that calculations of reserves and yields became somewhat more reliable.[55]

Today's research findings suggest that the notion of a general wood short-age on the eve of industrialization is not supported by the records.[56] The wood shortage only came into play for some social groups, or for industries that were dependent on particular types of wood, and for technical reasons could not fall back on other varieties. On the other hand, it is certain that wood stocks must have become more stretched by a growing population and increasing economic development. Thus, the declarations of a wood shortage anticipated to some extent a general crisis in supply. Is the warning of an imminent wood crisis therefore a false alarm? It is possible that this alarm contributed decisively to preventing an actual general shortage of wood. How important was the scare over wood shortages for the historical process?

THE EFFECTS OF THE ALARM OVER WOOD SHORTAGES: AN ATTEMPT AT A CLOSING APPRAISAL

It is difficult to prove any direct effect that the alarm over wood shortages in preindustrial society had on trade related to the forest, because such an alarm cannot be chronologically separated from the measures taken to combat the crisis. And even if one could prove the existence of a preindustrial environ-mental awareness avant la lettre, this would by no means signify that it was decisive in provoking action.

Prior to industrialization, it was obvious to everyone how important the forest and the resource of wood were to a society. We can see that there was general awareness of the finite character of natural resources from a 1773 competition organized by the Prussian Academy of Sciences on the topic: "What is the surest and easiest way to foster and to speed up the growth of trees in the forest, without the proposed method meaning that the wood loses firmness and strength, but on the contrary gains it?"[57] In the course of the eighteenth century it became a commonplace that the supplies of wood in the forests were by no means inexhaustible. It was also recognized in Berlin that one should search for means and ways by which to increase wood produc-tion. The reference to growth was indicative of economic thinking at the time; a solution was no longer sought only on the side of consumption, instead there was a desire to improve productivity and efficiency.

As a result of bottlenecks in transportation, in many areas people had already experienced a scarcity of wood, which usually manifested itself in the form of a rise in prices. Many forests were insufficiently developed in terms of transport infrastructure, meaning that only part of the usable wood resources in the forests actually reached the market.[58] However, until the nineteenth century, instances of scarcity were part of everyday life for the general population and for industry. People grappled again and again with shortages of food and raw materials, meaning that everyone had had their own experience of scarcity. Against this background, the warning of a wood shortage initially appeared as a new variation on a general lack. Nevertheless, the dreaded wood crisis had its own uniquely threatening quality because people knew that, by comparison, shortages of food were short-lived and could be remedied in the short term, for example, by grain imports. Other types of scarcity, for example a lack of rags to make paper, only affected particular specialized industries or social groups, leaving the majority unaffected. By contrast, a lack of wood threatened the whole of society and because of the slow growth rate of trees it could not be remedied within a few years. On top of that, the high cost of transporting wood hindered or prevented the gap from being plugged by other regions that were rich in forests. But the wood shortage was also particularly different from other scares in that no one questioned the necessity of preventing it. This meant that in situations of conflict, both the sovereign rulers and the territorial estates and communities could use this argument. Admittedly, there was no consensus about what solution should be pursued.

However, the imminent threat of shortages that was frequently referred to did not lead to a correspondingly careful treatment of the forest. In the Ancien Régime, the sovereign rulers' rhetoric about wood shortages often served to divert attention away from the massively exploitative logging that was going on in their own forests, and that had risen dramatically after 1750. The accusation that the forest was being devastated should in many cases have been turned back on the ruling princes, who, because they needed money, initiated a "drastic program of deforestation and cashing in on wood resources." While the damage done to young trees by the sovereign's game animals and by the hunts themselves was ignored, it was the peasants grazing animals and using wood who were blamed for the crisis. And it is true that in many cases they were hardly more careful with the forest than their lords were.[59]

The ruthless policies of exploitation that many ruling lords implemented came to an end with the eighteenth century, after which they were no longer evident in the German-speaking world, with the exception of pillaging in occupied foreign territory. With the new century, forestry underwent a professionalization, which had already manifested itself for the first time in the foundation of forestry academies. The foresters who had been trained in Tharandt, Neustadt-Eberswalde, Weihenstephan, Aschaffenburg, Hohenheim, or Giessen brought a new spirit to the forest. The main aim of forestry was no longer to get the highest possible economic return, but rather to produce the largest possible amount of wood in the long term. The principle of sustainability was at the top of forestry's hierarchy of values: future generations had to be able to use at least the same amount of wood as the current generation. In contrast to forest usage under the Ancien Régime, this principle was applied even at times when state finances were in trouble; in many states this was probably the result of secondary effects of the early constitutional system. Financial interests now came in second to the goal of getting the highest possible yield of wood, but they were still of great importance, in particular when it came to the utilization of wood. The protection of sustainable wood production had priority over all other claims, no matter whether these came from a fiscal authority, from the population, or from industry.

Because the foresters explicitly presented this new form of management as a set of necessary measures against the shortage of wood, the development of so-called rational forestry can actually be seen as a reaction to the eighteenth-century alarm over wood shortages. In this regard, Kurt Mantel's assertion that the wood shortage gave birth to forestry, quoted above, is accurate. In the historiography of the forest, the effect of this discourse and the reality of a wood shortage were often erroneously equated: it was not the shortage of wood itself, but rather the fear of it that led to the formation of scientifically based forestry.

From a silvicultural perspective, the nineteenth century was characterized by high forests of fully grown trees and by conifers. Since a fully grown forest produced the highest yearly yield of wood, this form of management was favored in most state forests. Numerous low- and medium-growing forests with shorter growth periods were converted into high forests in order to increase wood production. But this forest policy designed to prevent wood shortages had some unpleasant side effects:

1. Limiting the amount of logging in accordance with the ability of the forest to regenerate was an ecological and economic necessity. As a result, the provision of wood necessarily became worse for the poorer

sections of society, who could barely cover their basic needs anymore, what with limited supply and increasing demand, and who helped themselves by illegal means. In this way, they satisfied their demand, but often in areas of forest that badly needed to be conserved.

2. By lengthening the planned growth period of a forest, the foresters likewise postponed the time it would be used. This meant that every time a short rotation coppice was converted into a forest of full-grown trees, wood reserves were created for future generations. At the same time, these wood reserves became unavailable to the current generation for the foreseeable future and this reduced (for several decades) sustainably available supplies. The scenario that Joachim Radkau had predicted came to pass in the nineteenth century in the regions on the left bank of the Rhine: the impact of a wood shortage, which new forest policies sought to prevent from coming to pass in the future, was already being created in the present through a tightening of supply. Yet this was only an apparent shortage, because in fact demand was nevertheless being met, but this was happening through the black market on an unquantifiable scale.[60]

3. From an ecological perspective, the result of "rational forestry" was also not only positive. It is a historical achievement by the foresters that the forests were maintained despite the constantly growing demands on them. Since adhering to the principle of sustainability while also seeking the greatest possible yield of wood could only be calculated and controlled with relative certainty in areas of woodland that were consistent and homogeneous, the different areas of the forest were consolidated into stands and sections. These were subjected to homogeneous treatment, in which any troublesome elements (trees that were too old or too young, unwanted types of tree, tillering that was too dense or too sparse) were systematically removed. Now the forest, which had once been heterogeneously composed and rich in different species, had become a field of wood; the mixed forest had become a uniform forest of conifers. Insect damage and disease now afflicted the forest with renewed strength.[61]

Should we thank the alarm over wood shortages in the early modern period for the preservation and the form of today's forests? Such a simplified chain of cause and effect does not stand up to examination since many other factors have certainly strongly influenced forest structures, for example, bureaucratic calls for the forest to be organized in such a way that it could easily be monitored and controlled, or the long-term fiscal interest in having a steady high income from forestry.

However, forestry's effective commitment to the principle of sustainability would hardly have been conceivable and would never have been enforceable without the fear of a wood shortage. The wide dissemination and use of this argument was based on the fact that it was the universal consensus that a general shortage of wood absolutely had to be prevented. From the eighteenth century onward the population and the economy experienced strong growth, which entailed a growing demand for wood. Thus it was clear that one day this growth would necessarily meet its organic limits. Because the possibility of a wood shortage was articulated not only from "below" but also from the perspective of the sovereign rulers from "above," the alarm over wood shortages cannot be classified as a warning from the wilderness, or as the call of a Cassandra. Instead, the "wood shortage" was a general discourse, behind which a variety of interests related to the forest were hidden. When this discourse also gained political traction under German early constitutionalism and within the bureaucracy of reform, considerable obstacles stood in the way of reckless exploitation not just in the sovereign forests but also elsewhere, and woodland was protected from excessive use.

Finally, we should emphasize that prior to industrialization in Germany, wood shortages were a real phenomenon in a socially, temporally, and specially limited context. However, a general wood shortage that threatened the economic structure and the existence of society never came about. The alarm over wood shortages was an anticipated crisis; the general shortage of wood was not a reality. Nonetheless, the fear of it was justified. Yet the complaint that there was a wood shortage often became a false alarm when it was used as an argument against other interests competing for wood as a resource.

2

GRASSROOTS APOCALYPTICISM

THE GREAT UPCOMING AIR POLLUTION DISASTER
IN POSTWAR AMERICA

FRANK UEKÖTTER

IN JANUARY 1970, *LIFE MAGAZINE* offered dramatic news: "Horrors lie in wait." The magazine declared that according to expert opinions of the day, there was "solid experimental and theoretical evidence" for a number of scary prophecies. Ten years into the future, city dwellers would need gas masks to survive in the ambient air. By 1985, air pollution would block sunshine to such an extent that only 50 percent of solar radiation would reach the earth's surface. Rising carbon dioxide levels would change the world's climate, though *LIFE* was uncertain whether this would lead to a new ice age or to global warming. And then there was the great upcoming air pollution disaster: "In the early 1980s air pollution combined with a temperature inversion will kill thousands in some U.S. city."[1]

Gloomy environmental predictions were the order of the day in 1970. One month after the *LIFE* article, *Time Magazine* quoted ecology professor Kenneth Watt's warning of an upcoming smog disaster in Los Angeles: it would come within the next five years and cause mass death, "perhaps beginning in Long Beach."[2] In November 1970, the *Cleveland Press* called for dramatic steps against air pollution with another dark warning: "The alternative is the end of the human race, possibly in less than 30 years!"[3] Another Cleveland newspaper, the *Plain Dealer*, cited warnings of humanity's gradual suffocation for lack of oxygen, and future countermeasures would demand a high price: "We will be forced to sacrifice democracy by the laws that will protect us from further pollution."[4] It was obviously no exaggeration when a 1970 book, *The Politics of Pollution*, spoke of a "constant barrage of apocalyptic statements."[5] Even Willard Libby, recipient of the 1960 Nobel Prize in Chemistry and otherwise known as a staunch advocate of nuclear armament, declared in the summer of 1970: "A disaster due to hydrocarbons will occur in Los Angeles this fall."[6]

Environmental historians have long recognized the popularity of apocalyptic scenarios around 1970. More than a quarter century ago, John McCormick published a chapter on the environmental movement between 1968 and 1972 under the title: "The Prophets of Doom."[7] Apocalyptic scenarios defined the image of the nascent environmental movements, and statements from that time serve as lightning rods to the present day: the libertarian advocacy group FreedomWorks cites the *LIFE* article on its website as one of the "13 worst predictions made on Earth Day, 1970."[8] However, the days of gloom had a prehistory that has received far less attention among scholars and has left no mark in collective memory. In the years prior to 1970, people from widely different backgrounds resorted to alarmist language in complaints about air pollution with growing frequency, reflecting an increasing level of concern along with dismay that so little was done about it. It was not that reckless experts and journalists resorted to overblown prophecies in the years around 1970. It was merely that the antipollution discourse had previously developed in a way that allowed few other options for expressing concern. Apocalypticism prevailed in protests and rhetoric long before it left its mark in statements around 1970, and it frames understandings of the hazards of air pollution to this day. While horror scenarios capture the public imagination, chronic hazards remain an issue that rarely moves beyond expert circles.

The chapter explores the emergence of grassroots apocalypticism in four steps. The first section focuses on Los Angeles, the first U.S. city to experience photochemical smog. It shows that concerns about the health effects inspired gloomy predictions that had no precedent in American air pollution rhetoric. The second section shows how the rest of the country followed the path of Los Angeles in the 1960s: alarmist scenarios crept into the anti-pollution discourse and emerged as the common currency of pertaining discussions. The third section discusses the potential and actual consequences for American environmental policy. It shows that while the sense of alarm undoubtedly helped in the passage of strict anti-pollution legislation, some contemporary environmentalists had mixed feelings, as alarmist rhetoric could weaken and even jeopardize the drive against air pollution. The conclusion reflects on the legacy that the infatuation with disaster scenarios left for air pollution debates after 1970. Rather than flagging the chronic effects of air pollution, the environmental discourse sensitized for a dramatic event that never came.

MAKING SENSE OF A NEW THREAT: LOS ANGELES SMOG

In the United States, as in most industrial societies, coal smoke was the defining air pollution problem. While city dwellers lived and coughed in smoke and soot, urban reformers passed antismoke laws, hired engineers for enforcement, and thus helped to build a cadre of experts that shaped American air pollution policy into the postwar years.[9] However, Los Angeles was different. Abundant oil and gas defined California's energy regime, and the market share of smoke-emitting coal was less than 1 percent since the mid-1920s.[10] The absence of a major industrial sector helped, too: "As the city is not a great manufacturing center, smoke abatement is not a subject of very live interest," a smoke abatement survey of 1924 observed.[11] Pure air was one of the town's chief assets, and people suffering from tuberculosis played a significant role in the growth of Los Angeles in the late nineteenth century.[12] When John Anson Ford, a member of the Los Angeles County Board of Supervisors, opened the Mayor's Conference on Control of Smoke and Fumes in May 1946, he read a letter from a citizen suffering from asthma attacks who had recently come to Los Angeles from Seattle "in order to enjoy the sunshine, wonderful, clear mountain air and health-giving qualities of this area, as publicized so widely by the State of California."[13]

Los Angeles experienced its first smog episode in 1943. It was several years before the nature of the problem and the underlying causes were fully understood, and a groundswell of protest drove the efforts of scientists and officials: after decades of clean air, Los Angeles had lost its innocence.[14] Disaster scenarios became a significant part of the debate, and the people who put them forward came from many different backgrounds. For example, civic groups warned of an upcoming catastrophe when they made the case for stricter anti-pollution policies: "The increase in air pollution indicates a disaster is imminent," the Pasadena-based Citizens Anti-Smog Action Committee noted in a resolution of 1955.[15] A few years later, a resident of Hollywood declared in a letter to his California State Senator, "A terrible tragedy might well impend."[16] "While Californians exposed to smog complain most about eye irritation and other comparative minor effects, there is an underlying anxiety that the State may suffer . . . a sudden wave of deaths," the California State Department of Public Health wrote in its 1956 report "Clean Air for California."[17] It did not dare to play down pertinent fears: "We do not know how and when such a disaster could occur, because the determining factors responsible for such fatal episodes are not understood."[18]

People could point to a number of precedents where air pollution had grown so bad that it became a cause of imminent death. It had happened in the Belgium Meuse Valley in 1930, in Donora, Pennsylvania, in 1948, and in London in 1952.[19] The Meuse Valley fog was quickly forgotten, though a bronze statue and a plaque were put in place in 2000.[20] However, the disasters in Donora and London were part of living memory in smog-plagued Los Angeles. When James Roosevelt, the eldest son of Franklin and Eleanor Roosevelt, was running for governor in California in 1950, a campaign memo asserted, "Some day a great catastrophe will occur: thousands of persons will sicken and die as they did in Donora, Pennsylvania."[21] On "the problem of smog", the California-based radio station KPFA declared in a "Medical Care Broadcast" in 1956, "The danger of sudden mass disasters has already been demonstrated with tragic emphasis in the large-scale deaths in Donora, Pa. (1948), [and] in London (1952) when some 3000 persons were killed."[22] The event with the greatest similarity to Los Angeles occurred in Poza Rica, Mexico, in 1950. Just as Los Angeles did, Poza Rica had a large petroleum industry and was prone to inversion conditions. But as an event beyond the sphere of the West, the Poza Rica disaster did not qualify for much attention.[23]

The disinterest in the Poza Rica disaster sheds a revealing light on the significance of these air pollution disasters. The residents of Los Angeles saw them as symbols rather than precedents and never bothered to make a serious effort to study these events more closely. The London disaster was a particularly powerful symbol, as its death toll was staggering. According to contemporary estimates, 4,000 residents died prematurely; more recent estimates put the number as high as 12,000.[24] But London's curse was coal smoke, a marginal problem in Southern California. Donora was more similar to the situation in Los Angeles, if only because an inversion layer was among the causes of the 1948 disaster. But with pollutants coming from the local zinc works and Donora's location in the narrow Monongahela River valley, the situation differed markedly from Los Angeles and its photochemical smog.[25] Furthermore, with twenty people having died during the episode, Donora's death toll was far less dramatic—presumably the reason that the KPFA broadcast conflated the numbers into one. Moreover, whether Californians could identify with an industrial town more than two thousand miles away remains anyone's guess.[26] While the victims of Donora were mostly workers and their families, the struggle against Los Angeles smog was always dear to the heart of the affluent West Coast elite.[27] One of the most passionate champions of smog control was Stephen W. Royce, owner and general manager of Pasadena's Huntington Hotel, who rallied to the cause after several wealthy East Coast men told him that they would spend the next winter season elsewhere.[28]

In short, Donora and London were too different to serve as convincing precedents. They showed that air pollution could kill, but they would have put residents of smoke-stained cities and highly industrialized valleys on notice, and horror scenarios were conspicuously absent from discussions in these places during the 1950s. The likely source of Los Angeles apocalypticism lay within the people's bodily experience: residents felt the effects of photochemical smog in their eyes and lungs. "Upon arrival in Pasadena, I was compelled to put cold compresses over my eyes for ten to fifteen minutes before I could pay attention to close work," a car driver complained in 1950.[29] With watering eyes, burning throats, and cramping chests, Los Angeles residents were pondering the long-term health effects. As a resident declared in a letter to the Los Angeles County supervisor Kenneth Hahn in 1953, "If the smog is capable of eating microscopic holes in the clothes, which people wear, what damage must it be doing to your eyes, the delicate membranes of the nose, throat, bronchial tubes and lungs."[30] "This smog condition in the Pasadena area is just about unbearable. It has made everyone in my office ill, and has seriously affected the health of my wife and myself," another citizen wrote in 1946.[31] A medical doctor declared a few years later, "As a physician I see daily the deleterious effects of smog on my patients, especially those with chronic diseases of lungs and the heart. It is undoubtedly, to my way of thinking, shortening the lives of many of our citizens."[32] It was not a singular opinion, as the Los Angeles County Medical Association found out when it conducted a poll in 1950. The association "received the largest return of answers ever given any matter investigated by the association, whether for vote, or changes in policy, or otherwise. Nearly 70% of the association replied and over 95% of those replying stated that our 'smog' was not merely a nuisance but was a health menace, and the various systems involved were listed."[33] It was a shocking result, and not one that the association wanted to share with the wider public. Francis M. Pottenger, chairman of the Air Pollution Study Committee of the Los Angeles County Medical Association, described the association's reaction in a private letter: "We have not released these figures for fear that public panic might be created."[34]

Los Angeles smog was not the only environmental issue that provoked apocalyptic fears in the 1950s and early 1960s. Nuclear power provoked horror scenarios from the moment the first atomic bomb exploded in the New Mexico desert in 1945, and a transnational debate raged over fallout from nuclear testing after the United States conducted the Castle Bravo hydrogen bomb test on the Bikini Atoll in March 1954 and contaminated the Japanese fishing vessel *Lucky Dragon No. 5*.[35] Rachel Carson's *Silent Spring* famously began with "a fable for tomorrow" that described a town without singing birds.[36] But both horror scenarios hinged on an element of cultural construction: the looming danger required some imagination. Los Angeles smog was known to every breathing resident of Southern California, and it provoked dark fears among people of many different ways of life. Los Angeles apocalypticism grew from the grassroots, or perhaps more literally from the respiratory tract where smog claimed an unknown toll with every breath that people took.

For residents of Los Angeles, smog was a burning problem, and that provided air pollution control with an urgency that differed markedly from the more lackluster efforts in the rest of the country. As Kenneth Hahn declared in a letter to Gordon P. Larson, director of the Los Angeles County Air Pollution Control District in 1953, "There is no other problem facing our county that is more important than to restore healthy and clear atmosphere to our citizens, and I urge you to have a vigorous, all-out enforcement of existing laws so as to relieve the public of this serious smog nuisance."[37] Larson tried his best to follow suit, but when his efforts failed to restore the region's pristine atmosphere, he lost his job at the end of 1954. Everyone agreed that he had conducted a pathbreaking campaign, and nobody specified an act of negligence or incompetence that would have justified his resignation, but with public opinion at the boiling point, a rolling head was just to everyone's taste.[38] That made Los Angeles smog an ambiguous precedent when a similar mood captured the rest of the country in the 1960s.

FACING DISASTER: THE GREAT AIR POLLUTION SCARE

Los Angeles residents perceived photochemical smog as a health issue from the very beginning. The situation was different with respect to smoke and soot: they were usually seen as a threat to cleanliness and real estate values, with concerns about people's health figuring as a mere afterthought at best. "The medical argument against air pollution always was a hard sell," Adam Rome wrote about the ruling wisdom of the late nineteenth and early twentieth centuries.[39] But this negligence changed in the postwar years when contagious diseases lost much of their horror, and new long-term threats such as cancer gained prominence in popular perceptions of death and disease. As the Minneapolis-based Metro Clean Air Committee declared in 1970, "We should be very concerned by the rising incidence of respiratory disease including emphysema, asthma, chronic bronchitis, and lung cancer."[40] It was an important cause of growing concern about air pollution, which accelerated particularly in the late 1960s. The Opinion Research Corporation in Princeton, New Jersey, conducted a series of polls that showed how the share of respondents who saw air pollution problems as "very serious" or "somewhat serious" rose from 28 percent in May 1965 to 48 percent in November 1966 and 55 percent in November 1968.[41]

The growing urgency of air pollution protests also mirrored a sense of disaffection for the iron grip that industrial interests had on the approach to air pollution control. As Pittsburgh's widely acclaimed Allegheny County Bureau of Smoke Control declared in a review of its program, "Were it not for the cooperation of industry, this Bureau might as well turn out its lights and close its doors."[42] As late as 1967, the department's head declared with satisfaction that industry had entered "into a voluntary program in 1960 in which the steel industry agreed to equip all facilities with cleaning devices according to a 10-year schedule (1960–1970)."[43] This was worlds away from the situation in Los Angeles, where the County Air Pollution Control District received an angry letter from Kenneth Hahn when it gave refineries between eighteen and twenty-four months to carry out retrofitting in 1953.[44] It shows the dismal state of air pollution control in postwar America that the Los Angeles County

Air Pollution Control District alone claimed more than 40 percent of the total expenses for local control programs in the entire United States in 1961.[45]

Against this background, tempers rose in a way reminiscent of postwar Los Angeles, and just as in California, it is pointless to search for the original versions of horror scenarios. The apocalypse was everywhere, and it grew to a cacophonic crescendo around 1970. It was in the technical literature. "The capacity of our atmosphere is not limitless as many would like to think. At the present rate at which fossil fuels are being consumed, a few short years will suffice to render some of our cities uninhabitable," an article in *Air Conditioning, Heating and Ventilation* declared.[46] The *Harvard Business Review* quoted the warning of meteorologist Morris Neiburger in 1966: "The world's atmosphere will grow more and more polluted until, a century from now, it will be too poisonous to allow human life to survive, and civilization will pass away."[47] An advertising newsletter of the Philadelphia Gas Works stated: "If air pollution problems are not solved soon, lunch pails some day may contain a can of fresh air rather than soft drink, and lunch counters may serve a can of cool, fresh air instead of water."[48] A brochure of the American Federation of Labor and Congress of Industrial Organizations (AFL-CIO) of 1969 asserted: "When the right circumstances conspire, air pollution can turn into a deadly mass killer."[49] And a set of poems from a third-grade creative writing project at St. John Lutheran School in Cleveland told the city's mayor, "Soon we will walk around in gas masks."[50]

With so many different people expressing similar fears, it is not surprising that dark warnings became a defining part of the political discourse. For example, a circular of the Conservation Foundation declared in 1966, "Some experts warn that as many as 10,000 people may die prematurely in the near future in one of the largest cities of the world which are blanketed by smog." The Conservation Foundation even offered some "leading candidates for such a disaster . . . New York City; Los Angeles; London; Santiago, Chile; and Hamburg, Germany."[51] The Air Pollution Control Commissioner of New York City, Leonard Greenburg, used horror scenarios to boost his political standing: "The lack of recognition of the importance of this problem and the appropriation of insufficient funds may well mean that 10 [years] from now New York will find itself in the position in which the city of London found itself when more than 4,000 deaths were brought about by air pollution."[52] In 1963, a staff report of the Senate Committee on Public Works warned that air quality might deteriorate "to the point where episodes of acute illness and even deaths [are] more than occasional."[53] In a society where disaster scenarios struck a nerve, it was illusionary to expect more caution from political figures, and yet one cannot help but wonder whether they were harboring second thoughts. After all, prophecies of upcoming disasters could wear thin after a while if they did not materialize, and activists certainly had other options to make their case. There was a significant body of evidence for the deleterious effects of air pollution short of sudden mass death.

It does not seem that people were reflecting extensively on the pros and cons of environmental apocalypticism in the 1960s, but there is ample evidence that they took the possibility of a major disaster seriously. It is remarkable that environmentalists even followed up on outlandish scenarios. For example, when rumors were circulating that the earth's atmosphere might run out of oxygen, the National Audubon Society contacted a geographer at the University of California, Berkeley, who told the association that "consumption of the oxygen is the least of several worries; greater ones are the effects of a possible doubling of the carbon dioxide content of the air and the . . . flooding of the world's lowlands."[54] However, it is revealing that the Public Health Service published a booklet in the mid-1960s featuring selected cartoons on air pollution. Dramatization is a legitimate part of the cartoon business, and the booklet offered a generous dose of dark humor: one drawing showed a businessman standing on the brink of a devastated river landscape with his son proclaiming, "Someday, my boy, this will all be yours."[55] According to its title, the booklet was "no laughing matter," and yet the federal agency obviously felt that it was good idea to distribute apocalyptic imagery tongue in cheek. The Public Health Service was not serious about any of the warnings that the cartoonists had sketched. It was serious about pollution.

With that, one should not regard the fear of the great upcoming air pollution disaster as a phenomenon in its own right. Horror scenarios reflected a genuine fear that captured a growing part of the American population in the 1960s. As more and more people were concerned about pollution, the general thinking was that things were bad enough and that making them look even worse could not do any harm. Characteristically, the upcoming disaster remained notoriously diffuse, with many different scripts and places in play and a plethora of parties who kept them circulating: the common denominator of the horror scenarios was that they were scary, somewhere in the not-too-distant future, and that they involved air pollution. In any case, the important thing was not the specific scenario but that somebody would do something about the problem soon. It was a significant boost to policymaking, but as it turned out, horror scenarios had ambiguities in the political sphere. The path from the apocalypse to effective policy was not a one-way street.

DEALING WITH IT: POLICYMAKING AND THE APOCALYPSE

Historians agree about the significance of 1970 for American environmentalism. Congress passed landmark legislation during that year, the Nixon administration sought to brush up its ecological credentials, and an estimated twenty million people celebrated Earth Day on April 22, 1970.[56] Air pollution was among the signature issues of protest and policymaking, and the Clean Air Act of 1970 continues to shape U.S. air pollution control to the present day. It was a genuine outpouring of environmental concern, and apocalyptic rhetoric was the language of the day. In a letter written on the occasion of Earth Day, the Minnesota Emergency Conservation Committee told Bethlehem Steel: "Because of the conditions that your industry and other industries have acted towards your fellowmen and towards your nation this once great nation faces oblivion."[57] The August 1971 issue of *Playboy* featured a cartoon that showed a job-seeking youngster facing a manager against the backdrop of the factory's devastating pollution, with the manager cheerily announcing a company policy: "We don't have a retirement plan. We don't think the country will last that long."[58]

But for the first time, some skeptical remarks mixed into the groundswell of alarmist statements. Robert Moses, the legendary urban planner of New York City, lambasted the sense of panic in the September 1970 issue of *National Review*, "The four pale horses of the Apocalypse are back again, this time with a wealth of scientific guff calculated to force conservation by terror instead of promoting it by logic, honest arithmetic, patience and sacrifice."[59] Around the same time, *Time* published an article on "the rise of anti-ecology" that, while generally sympathetic to the cause of environmentalism, took issue with the diffuse nature of grassroots apocalypticism: "The key problem seems to be that the rhetoric of ecology too often makes the subject look like a confused mix of unrelated alarms and issues."[60] Even activists were having second thoughts, particularly those who took pride in their own professionalism. In a pamphlet titled "How to Testify at a Public Hearing," the Pittsburgh-based Allegheny County Environmental Coalition warned, "Emotional statements can often do more harm than good. Your valid points can be lost if some of your testimony is misleading or irrational. Document your facts and be sure that what you say is true. Don't tell them the sky is falling unless you have a piece of it in your pocket."[61]

Environmental alarmism could even backfire, as the Sierra Club learned in 1969 when a Hollywood-based environmental group named People's Lobby Inc. came up with a ballot initiative for California.[62] Thanks to Los Angeles smog, the state had plenty of experience in fighting air pollution as well as an established network of laws and institutions, and it soon became clear that the initiative was amateurish and potentially damaging to existing policies. "The Initiative . . . might seriously hinder efforts to solve the air pollution problem," the Sierra Club warned. The initiative pursued criminal charges against polluters, but language was "so vague as to probably make enforcement unconstitutional."[63] The initiative sought to repeal existing legislation, thus eroding the legal foundation on which ongoing antipollution policies were based. There was also a chance "that the State of California will lose its special status under federal law"—the state had successfully fought for an exemption from federal moving vehicle air pollution legislation in order to impose stricter standards for Los Angeles.[64] Other experts shared these concerns. "The initiative proposal would create problems of uncertainty and vagueness," the Legislative Counsel of California declared, and the Los Angeles Office of the County Counsel, well-versed in the juridical details of the country's toughest air pollution control program, urged the Board of Supervisors of Los Angeles County to oppose the initiative: "The measures would not be effectual and would destroy the existing structure of air pollution control in the State of California."[65] But when the Sierra Club came out against the initiative, it fought an uphill battle and struggled to dispel charges of "organizational conservatism." In the face of an initiative that came across as "the only hope for reducing air pollution," the Sierra Club was stunned to discover "that people are not convinced . . . that in fact we are trying to do something about the problem."[66]

The episode showed the perils of policymaking by sentiment. Grassroots apocalypticism was a powerful force when it supported informed initiatives but a major irritant if it seized on poorly conceived efforts. The sense of alarm put a premium on determination and discounted technical and juridical expertise, and if it had not been for the latter, the environmental initiative of People's Lobby might have inadvertently wrecked Californian air pollution control. In other words, the outburst of environmental sentiments could well have led to meager or even counterproductive results if it had not been for a preexisting network of activists, officials, and policybrokers who used the moment to their advantage. After all, the heydays of environmentalism were fading into memory with amazing speed. By the end of 1972, the executive committee minutes of the Portland-based Oregon/Washington Coalition for Clean Air in Oregon und Washington recorded a change of tide: "It was felt that many environmentalists were becoming apathetic about air pollution because many of the major issues have been 'technically' solved in Oregon, i.e., legislation or other effective means to solve the problems are available." The coalition found this to be a depressing outlook, for everything ultimately relied on stringent enforcement: "The good work of the past could be lost by non-implementation or by the steady erosion of contrary legislation or increasing variances from standards."[67] The apocalypse was less than helpful when it came to the long haul.

CONCLUSION: FACING CHRONIC THREATS

Gas masks remain optional in American cities, and the great air pollution disaster never materialized, but that did not end the environmentalists' infatuation with gloomy predictions, as anyone who has not slept through the climate change debate will readily attest. Alarmism lingers as the environmentalists' rhetoric of choice, and certainly not for lack of alternatives. A 2014 news release of the World Health Organization declared that about 7 million people died prematurely in 2012 because of exposure to air pollution, including 4.3 million deaths due to households that use coal, wood, or biomass for cooking.[68] However, the known effects of particulate emissions usually generate less excitement than the potential effects of changing climates in the future, at least among Western environmentalists. All the while, the ambiguous political merits of horror scenarios have become evident on the world stage in the form of the 2009 Copenhagen Climate Summit. It took place against the backdrop of apocalyptic scenarios, courtesy of the Fourth Assessment Report of the Intergovernmental Panel on Climate Change, Hurricane Katrina, and Roland Emmerich's blockbuster movie *The Day After Tomorrow*. Nonetheless, Copenhagen failed in spectacular fashion.

Some people already recognized in the 1960s that society's penchant for dramatic disaster scenarios was missing the real story. "The mortality associated with spectacular but temporary air pollution conditions is actually of less significance than a growing body of evidence that long-term, low-level air pollution contributes to and aggravates certain diseases," an article in *Editorial Research Reports* stated in 1964.[69] The chronic effects of air pollution remained a topic for expert hearings and never captured the public imagination in quite the same way as the future need for gas masks. Disasters were scary, easy to understand, and impossible to disprove, at least for the time being. Figures about premature deaths were statistics.

Perhaps the most striking parallelism between the air pollution discourse of the 1950s and 1960s and the climate change discourse of our time is the striking diffuseness of the apocalypse. The sense of looming disaster is widespread, but its exact nature takes a plethora of different shapes. Perhaps we have grown so accustomed to environmental disaster scenarios that we no longer care about what they seize upon? One of the climactic moments in Al Gore's 2006 documentary film, *An Inconvenient Truth*, was the prediction that rising sea levels would inundate the World Trade Center memorial in New York City. Perhaps future scholars will find this a fitting finale to the age of environmental apocalypticism: easy to grasp, powerful in symbolism, and yet without a connection to people's real lives. Stephen W. Royce certainly cared about the air around his Huntington Hotel in Pasadena. But who really cares about New York real estate?

3
"A COMPUTER'S VISION OF DOOMSDAY"

ON THE HISTORY OF THE 1972 STUDY
THE LIMITS TO GROWTH

PATRICK KUPPER AND ELKE SEEFRIED

IN THE SPRING OF 1972, humankind faced a grim future. The world was heading directly for the collapse of civilization—or it was at least if you believed *The Limits to Growth*, a publication commissioned by the Club of Rome. And many people did believe it. The tremendous amount of attention that this report attracted around the world and the frenzy of activity that it provoked in all quarters point to the fact that this slim volume must have tapped into a central preoccupation of its time.[1]

What was this preoccupation? What was *The Limits to Growth* really about? Who were the authors of the report, who commissioned it, and what was its purpose? What data was the report based on, what methods were used to interpret it, and how did it present the results? Was the study grounded in serious science or sensationalist quackery? And finally, what was the effect of the study, and why did this apocalyptic warning provoke such a huge social reaction worldwide?

To answer these questions, it is necessary to reconstruct the origins of *The Limits to Growth* and to name the most important players in its genesis, to examine the report itself and its effects, report its conclusions, and explain its ongoing resonance.

THE CLUB OF ROME AND THE ORIGINS
OF *THE LIMITS TO GROWTH*

The report was commissioned by the Club of Rome. This organization was founded in 1968 on the sidelines of a conference in Rome's Accademia dei Lincei. Its initiators were Aurelio Peccei, an Italian industrialist, and Alexander King, a Scot who was director of Scientific Affairs at the Organization for Economic Cooperation and Development (OECD). The economist Peccei had worked in South America as a manager for Fiat in the 1950s, then became managing director of Olivetti and the founder of Italconsult, a consultancy firm that oversaw development projects. Partly on the basis of these experiences, Peccei's goal became, initially in a thoroughly technocratic sense, the "modernization of society" on a global scale.[2] Peccei's starting point was the diagnosis that technological progress was accelerating. According to Peccei, modern technologies meant that developed, industrial societies were changing increasingly quickly and this was widening the gap between industrialized and developing countries. Likewise, technological acceleration was becoming a problem for Western Europe: here, he referenced a concept that was widespread in the 1960s of a growing "technological gap" between Western Europe and the United States.[3] Peccei therefore called for future global problems to be researched in greater detail with the aid of new technological means and to be dealt with through "world planning"—albeit under Western leadership. For Peccei was closely bound up with the Western liberal consensus, the American Democrats, the Ford Foundation, and the OECD.[4]

In fact, the OECD was another starting point for the Club of Rome in terms of organization and content. In the light of the debate over a "technological gap" between Western Europe and the United States, the OECD had—with Alexander King—commissioned a study in the mid-1960s into "technological forecasting," which was supposed to identify future technological potential in Western Europe and the ways in which it could be investigated.[5] Peccei and King met each other in 1967 in Paris. United by the belief that in the face of accelerating technological development it was necessary to improve methods of forecasting and long-term political planning, they founded the Club of Rome along with a group of intellectuals, scientists, and business representatives (first European, then also American and Japanese).[6]

"The Club of Rome is an informal, multinational, non political group of scientists, economists, planners, educators and business leaders," read an exposé of the club from 1970.[7] In fact, it was a highly elite, male-dominated, Western-influenced association that explicitly saw itself as being made up of the world's front-runners. Its members were carefully selected. New members were appointed by the governing Executive Committee, and membership was limited to a maximum of one hundred people. The Club of Rome attempted to ensure its political independence by stipulating that membership was incompatible with political office.[8] At the time *The Limits to Growth* was published, the Club of Rome had some seventy members from twenty-five countries.[9]

The Club of Rome proceeded from the assumption that the world had arrived at a turning point. It diagnosed a "Predicament of Mankind,"[10] the cause of which was identified, in accordance with Peccei, as the rapid technological and economic progress of recent decades. This progress was perceived as one-sided. Ethics, morals, ideals, and institutions had not kept pace, which had led to an imbalance. On the one hand, this progress had brought tremendous gains in prosperity, at least for people in industrialized countries, but on the other hand, it had also created a multitude of problems, the extent and, moreover, the interdependencies of which had not yet been grasped. Problems such as overpopulation, malnutrition, poverty, and environmental pollution should no longer be looked at in isolation, but instead had to be understood—and this was the very core of the far-reaching diagnosis—as integral components of a complex, all-encompassing worldwide problem that was labeled the "problématique": "It seems reasonable, therefore, to postulate that the fragmentation of reality into closed and well-bounded problems creates a new problem whose solution is clearly beyond the scope of the concepts we customarily employ. It is this generalized meta-problem (or *meta-system* of problems) which we have called and shall continue to call the "problématique" that characterizes our situation."[11] The Club of Rome had thus discernibly intensified Peccei's diagnosis, and named coming global problems that should be examined as systematically as possible, determining the dynamics of the interactions, "which seemingly exacerbate the situation as a whole."[12] These questions were debated at a conference, officially organized by the OECD, which took place in Bellagio in the fall of 1968. At this conference the significance of studies based on systems analysis for the wider work of the Club of Rome became clearly apparent.[13]

Systems analysis had its roots in World War II operations research, with which the American and British armies had conducted logistical and statistical optimization analyses for weapons and equipment systems.[14] In the emerging Cold War, and in the context of U.S. big science, operations research knowledge fused with cybernetics, developed as a new meta-science of communication and control in animate and inanimate systems, to form a new field of prediction, or "Futures Research." At think tanks such as the RAND Corporation and at the renowned Massachusetts Institute of Technology (MIT), systems analysis was conceptualized as a cybernetics-inspired "science of strategy,"[5] addressing the "complex problem of choice among alternative future systems, where the degrees of freedom and the uncertainties are large."[6] Systems analysis also served as an umbrella for forecasting methods such as computer-based simulation modeling. Computers seemed to offer the opportunity to collect data and to simulate the complex processes to be expected of an intricate system in analysis runs.[7] For the Club of Rome, systems analysis offered a perfect way of researching the worldwide "problématique" that needed to be examined in its systemic components and interactions. At the same time, systems analysis itself contributed to the perception of crisis as the "problématique" also arose from an assumption of increasing complexity and all-encompassing interaction of problems that epistemologically characterized systems analysis. Thus, the overinflated confidence in systems analysis was a contributor to the construction of a coming crisis. For the Club of Rome, the problématique seemed as if it could only be brought under control by long-term planning on a global scale. Thus, the club assumed the role of diffusing the findings of these investigations among the public and among political decision makers.[8]

In 1969 the Club of Rome decided to commission an initial project called "The Predicament of Mankind." The aim of the project was "to clarify to others as well as ourselves how the systemic nature of society's and the world's problems and therefore the critical interdependencies and interactions among them . . . can all be investigated, understood and described in such a way as to provide a basis—which is presently lacking—for the decision centers rationally and effectively to define goals, policies and strategies."[9]

The American economist and futures researcher Hasan Ozbekhan, who was himself a member of the Executive Committee of the Club of Rome, was at first entrusted with carrying out the project. However at an initial plenary session, which took place in Bern in June 1970 at the invitation of the Swiss government, the ambitious concept that Ozbekhan presented failed to convince club members. Ozbekhan had delivered a highly differentiated, intricately conceived paper on the possibilities of a systemic assessment of the world's problems.[20] However, the Executive Committee thought it was more important to be able to present a study for the general public soon and thus also generate a political influence. Moreover, the German professor of mechanics at the University of Hannover, Eduard Pestel, had made an effort to secure funding from the Volkswagen Foundation (which was also based in Hannover); he told the others in Bern that the foundation had asked that the report be modified.[21] Into the breach that opened up at the conference in Bern stepped Jay W. Forrester. The trained electrical engineer had worked in the Second World War in the field of operations research. After 1945, he developed the computerized radar defense system SAGE for the U.S. Air Force, which he embedded in a broad research and development program at MIT, through which he took a leading role in both the expansion of operations research into cybernetically designed systems analysis and the development of digital supercomputers.[22] In the 1960s he began to simulate complex developments (for example in companies) with computerized modeling and thus became a pioneer in constructing the already mentioned computer-based simulation models.[23] In Bern, Forrester proposed to the Club of Rome that he further develop his approach, which he called "System Dynamics," according to the club's need for a "World 1 model."

Consequently, the club's Executive Committee decided to hand the project over to MIT.[24] Leadership of the project passed to a young economist, Dennis Meadows, who was a student of Forrester's. Forrester himself did not take part in the project, but he did develop the methodological principles on which the work of the Meadows group was grounded. Forrester designed the "World 2 model" that formed the basis of his study *World Dynamics*, which he published in 1971, a year before *The Limits to Growth* came out.[25] Ozbekhan, by contrast, withdrew, whereby the project also limited itself to a purely quantitative approach. The Volkswagen Foundation provided financial support.[26] Under Meadows's leadership, the research team of seventeen people, made up chiefly of young researchers from different disciplines and of different nationalities, immediately got down to work in constructing the "World 3 model." In March 1972, after just one and a half years of work, Meadows and the club's Executive Committee were able to present the report, *The Limits to Growth*, to the public in the presence of senior figures from American and international politics and business (such as the under-secretary-general of the United Nations, Philippe de Seynes) at the Smithsonian Institute in Washington, DC.[27] In the same year, the report was translated into twelve languages, including French, German, Spanish, Italian, and Japanese.[28] The report was short and pithy, strongly results-oriented, and written in language that could be understood by a general audience. Underlying data and mathematical formulas were published one year later in a collected papers report.[29]

GROWTH TO THE DEATH? THE REPORT AND ITS CONSEQUENCES

Humanity is in the midst of digging its own grave. This was the easily comprehensible core message of *The Limits to Growth*. "If the present growth trends in world population, industrialization, pollution, food production, and resource depletion continue unchanged, the limits to growth on this planet will be reached sometime within the next one hundred years. The most probable result will be a rather sudden and uncontrollable decline in both population and industrial capacity."[30] It still seemed "possible to alter these growth trends and to establish a condition of ecological and economic stability"[31]:

> The way to proceed is clear. . . . Man possesses, for a small moment in his history, the most powerful combination of knowledge, tools, and resources the world has ever known. He has all that is physically necessary to create a totally new form of human society—one that would be built to last for generations. The two missing ingredients are a realistic, long-term goal that can guide mankind to the equilibrium society and the human will to achieve that goal. Without such a goal and a commitment to it, short-term concerns will generate the exponential growth that drives the world system toward the limits of the earth and ultimate collapse.[32]

In its critical appraisal, which was appended to the end of the report, the Executive Committee of the Club of Rome called for nothing less than "the initiation of new forms of thinking that will lead to a fundamental revision of human behavior and, by implication, of the entire fabric of present-day society."[33]

How did Meadows and his team arrive at its alarming statement? Electronic data processing, systems analysis, and the construction of a world model were the key approaches through which the researchers at MIT sought to grasp reality. "Ours is a formal, written model of the world. It

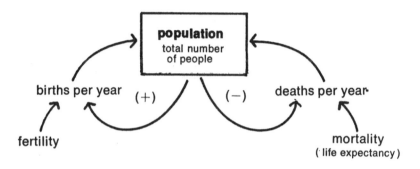

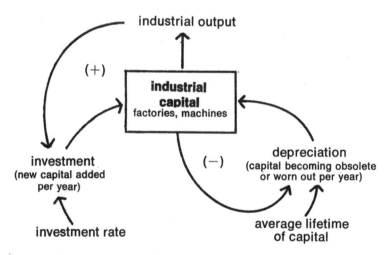

The central feedback loops of the world model govern the growth of population and of industrial capital. The two positive feedback loops involving
births and investment generate the exponential growth behavior of population and capital. The two negative feedback loops involving deaths and
·depreciation tend to regulate this exponential growth. The relative strengths
of the various loops depend on many other factors in the world system.

FIGURE 3.1. Populations growth and capital growth feedback loops. From
Meadows et al., *Limits*, 95.

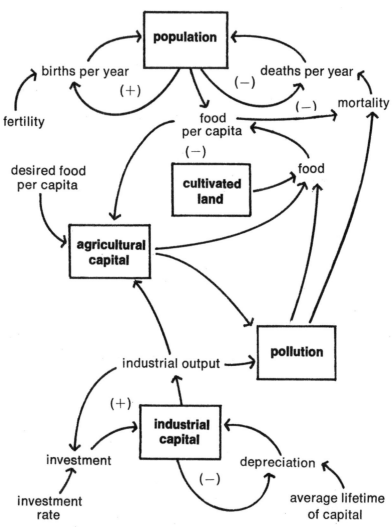

Some of the interconnections between population and industrial capital operate through agricultural capital, cultivated land, and pollution. Each arrow indicates a causal relationship, which may be immediate or delayed, large or small, positive or negative, depending on the assumptions included in each model run.

FIGURE 3.2. Feedback loops of population, capital, agriculture, and pollution. From Meadows et al., *Limits*, 97.

constitutes a preliminary attempt to improve our mental models of long-term, global problems by combining the large amount of information that is already in human minds and in written records with the new information-processing tools that mankind's increasing knowledge has produced—the scientific method, systems analysis, and the modern computer."[34] Five macroeconomic variables were singled out as being decisive for the "world system"—"population, food production, industrialization, pollution, and consumption of nonrenewable natural resources."[35] The interaction between these parameters was described in feedback loops, whereby the relationship between two variables could be either positive (i.e., mutually reinforcing) or negative. Altogether, the world model consisted of ninety-nine quantified variables, which were bound together in a "hundred or so causal links" (see figures 3.1 and 3.2).[36]

The second step was to supply the "world model" with quantitative data from all over the world. In a third and final step, the MIT team let the computer calculate, according to varying base assumptions, several runs of the model up to the year 2100 (see figures 3.3 and 3.4). The "Standard Run," that is, the iteration of the model that assumed that no major physical or political upheavals were to take place, resulted in a scenario according to which, in light of exponential economic and population growth, the limits to growth on earth would be reached before the year 2100. Rapidly growing industrial capital would require more and more raw materials and would pollute the environment. The price of raw materials would rise significantly with their increasing use, with the result that more and more investment would be necessary. At some point the industrial development would come to a standstill and as a result the population would also collapse. The team of authors compiled various alternative scenarios, which allowed for variations, for example, a doubling of raw material reserves, a more efficient use of resources, and a partial replacement of fossil fuels by nuclear energy, as well as an increase in agricultural yields with the aid of new technologies. Here too there loomed a future collapse, above all from increasing pollution.[37] Admittedly, the study was designed in such a way that it yoked together exponentially growing factors (population and industrial production) with fundamentally limited factors of global "carrying capacity" (food and nonrenewable resources) in a "finite world"[38]—and therefore the model inevitably pointed toward global collapse.[39]

However, the authors saw the formalization and quantification of applied modeling as having two important advantages over qualitative models of thought: "First, every assumption we make is written in a precise form so that it is open to inspection and criticism by all. Second, after the assumptions have been scrutinized, discussed, and revised to agree with our best current knowledge, their implications for the future behavior of the world system can be traced without error by a computer, no matter how complicated they become."[40] To be sure, in another passage the authors pointed out that their study dealt with scenarios and, therefore, "not exact predictions."[41] However, the adjectives "precise," "exact," and "accurate" revealed both the technocratic approach of the report's creators and their fascination with their own methods.

From the model, the MIT team deduced that swift action had to be taken. The "short doubling times of many of man's activities, combined with the immense quantities being doubled" meant that the limits to growth would be reached very soon. "Population and industrial capital reach levels high enough to create food and resource shortages before the year 2100."[42] The alarming message was that without further delay the course had to be altered and growth had to be consciously limited. For every day that went by without action being taken, the global system was propelled closer to its limits, and beyond, since: "The behavior mode of the system . . . is clearly that of overshoot and collapse."[43] In a way very similar to Forrester's in his earlier study, the team of authors around Meadows juxtaposed the looming apocalypse with the solution of "global equilibrium." Citing John Stuart Mill, the group declared a "controlled end to growth" and thus a "stationary condition of capital and population" as an idealized guiding principle. This implied authoritarian measures and limits to freedoms, for example, in making decisions about the number of children or the use of raw materials. The authors sketched out a postmaterial world, in which human development would still be possible, for instance, through placing a new emphasis on nonmaterial values, education, or art.[44] Consequently the MIT team and the Club of Rome made an appeal for the world to be reorganized through austerity and global planning. Thus, the study had a fundamental technocratic nature that came out of the original intentions of the Club of Rome founders as well as the computer model.

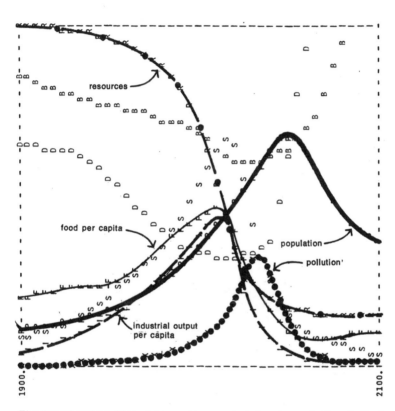

The "standard" world model run assumes no major change in the physical, economic, or social relationships that have historically governed the development of the world system. All variables plotted here follow historical values from 1900 to 1970. Food, industrial output, and population grow exponentially until the rapidly diminishing resource base forces a slowdown in industrial growth. Because of natural delays in the system, both population and pollution continue to increase for some time after the peak of industrialization. Population growth is finally halted by a rise in the death rate due to decreased food and medical services.

FIGURE 3.3. World model standard run. From Meadows et al., *Limits*, 124. Note that the letters in the figure are defined as follows: B is birthrate; D is death rate; and S is services per capita.

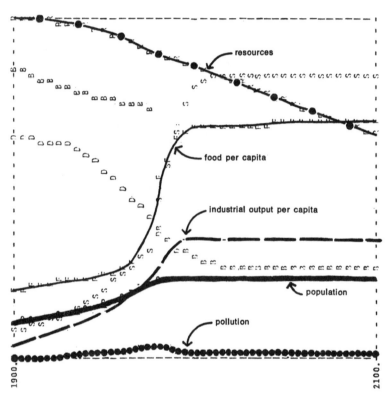

Technological policies are added to the growth-regulating policies of the previous run to produce an equilibrium state sustainable far into the future. Technological policies include resource recycling, pollution control devices, increased lifetime of all forms of capital, and methods to restore eroded and infertile soil. Value changes include increased emphasis on food and services rather than on industrial production. As in figure 45, births are set equal to deaths and industrial capital investment equal to capital depreciation. Equilibrium value of industrial output per capita is three times the 1970 world average.

FIGURE 3.4. Stabilized world model I. From Meadows et al., *Limits*, 165. Note that the letters in the figure are defined as follows: B is birthrate; D is death rate; and S is services per capita.

THE CONTEMPORARY DEBATE ABOUT *THE LIMITS TO GROWTH* AND THE EFFECTS OF THE REPORT

The public response to the publication of the study was absolutely extraordinary. Neither the Club of Rome, nor Dennis Meadows were known to the wider public before *The Limits to Growth*. Now all of a sudden they were world famous. "Almost never before has a mathematical expression of societal contexts created such a sensation as the world model of the Club of Rome," declared the economist Bruno S. Frey in the *Neue Zürcher Zeitung* in August 1972.[45] The book unleashed worldwide debates that were not soon put to rest. Influential media outlets published extensive reviews of the report and opened up discussion forums.[46] In the following months numerous publications appeared that explicitly discussed the report.[47] In 1973 the Club of Rome was even awarded the prestigious Friedenspreis des Deutschen Buchhandels (Peace Prize of the German Book Trade) for *The Limits to Growth*. The book was translated into thirty-seven languages and has sold more than twelve million copies to date.[48]

The reaction to the publication of the report was highly contentious and ranged from deep consternation to fierce criticism and outright rejection. The first newspaper reviews varied greatly in their assessments. The British *Guardian* wrote: "Documents that may change the course of mankind are very rare. 'The Limits to Growth' . . . is such a document."[49] The *New York Times* columnist Anthony Lewis saw the book as "one of the most important documents of our age" because it showed that "the elements of life are interconnected."[50] By contrast, for the *Economist* the report reached "the highwater mark of old fashioned nonsense," and *Foreign Affairs* wrote under the headline "The Computer That Printed Out W*O*L*F*": "The authors' analysis is gravely deficient and many of their strongest and most striking conclusions unwarranted." With a similarly damning review, titled "A Computer's Vision of Doomsday," the German news magazine the *Spiegel* joined the ranks of the critics after the publication of the German edition. Economic growth, not general stagnation, was necessary for the "preservation of the species," stated the magazine's ironically phrased conclusion.[51]

Criticism of the report came, broadly speaking, from three quarters. In the first place, the report was harshly criticized by prestigious economists and scholars of an empirical branch in futures research.[52] They objected that the data pool and the system of variables were far too small for the far-reaching assertions. What is more, they said that the economic assumptions at the basis of the study were erroneous. In particular, the study did not pay sufficient attention to pricing mechanisms or technological progress. For example, according to Paul A. Samuelson, Meadows was essentially making the same mistakes that Thomas Robert Malthus had made almost two hundred years earlier in his predictions about population growth.[53]

> When raw materials become scarce and disruptions to supply begin to suppress the growth rate, as in Forrester's simulations, then this will lead, in reality, to a rise in prices. In England, one used to burn wood in order to smelt iron. In the long term, this wasn't possible because England's woodland reserves suffered too much damage. Therefore, wood was dropped in favor of coal. This is also the great mistake that T. Robert Malthus made in his prophecy of a population catastrophe, published in 1798. Malthus did not predict the marvels of the industrial revolution. And the marvels of the industrial revolution have not yet come to an end.[54]

In this way, a Malthusian interpretation was confronted with a Cornucopian one. These were diametrically opposed, in particular in their assessment of technological progress and the question of how far this progress, supported by market-based price setting, would solve environmental problems or would make them even worse.[55] This was bound up with the issue of resources that were yet to be discovered: many economists criticized the fact that *The Limits to Growth* only considered resource reserves that were already known, whereas new resources were constantly being identified and new methods of extraction could be developed (the MIT team countered that even new discoveries would only delay the squeezes on resources, not prevent them[56]). In this group of critics, technological optimism was united with an admiration for the human individual and his capabilities. Thus, the empirically focused British research group Social and Technological Alternatives for the Future at the Science Policy Research Unit at the University of Sussex criticized the "technocratic" computer model for negating the individual and his potential.[57] Futures research, it was argued, had to devote more attention to people, their values and needs. Different futures were still fundamentally open: "Forecasting . . . cannot predict history to come, it is limited to the identification of possible futures and of problems that might have to be faced on the way to such futures."[58]

Second, the report also received harsh criticism from sections of the Left. Intellectuals and researchers questioned in particular a "world model" that neglected almost all political and social issues. Moreover, the bringer of these bad environmental tidings created suspicion. Leftists such as the German intellectual Hans Magnus Enzensberger or the Norwegian peace and futures researcher Johan Galtung argued that the working class had suffered the effects of environmental pollution since the beginning of industrialization and, in the poorest parts of the world, overpopulation and the overexploitation of resources by countries to the north had been evident for some time. The Club of Rome's catastrophic scenario was only being expressed now "because some of the same conditions have recently, due to some changes in the organization of production and in the technology, also reached well into the middle classes." In this way, ecology morphed into "middle-class ideology," Galtung declared.[59] Enzensberger shared this opinion and added, "If avowed representatives of monopoly capitalism have recently become its spokesmen—as in the Club of Rome," then the "monopolies" attempted to acquire influence over the ecological movement because of material interests, namely, those of the "eco-industrial complex." According to this thinking, the recognition of the problems attendant on industrial growth served to promote a new eco-industrial growth industry led by a few international concerns whereas increasing pollution had to be contained by control techniques financed by the public.[60] Further, Enzensberger pointed to the vital interests of power politics. The "captains of industry, gathered together in the Club of Rome, appear to have another view of conditions on board the ship in which we are supposed to be sitting. . . . This minority leaves no doubt that they are determined to adjust their view of the world to suit their own interests. The scarcer the resources the more one has to take this view in distributing them; but the more one adopts this view of the world the fewer people can be considered for this high office."[61] In the fatuous call for a radical U-turn, Enzensberger saw a depoliticization of the ecological question. In his opinion, not ecological factors, but social variables, not industrialization, but capitalism stood at the center of the environmental crisis.[62]

Third, related to this, *The Limits to Growth* was attacked because of its lack of regional differentiation. Scientists from threshold and developing countries as well as western experts who, from a critical perspective, called for a leveling out of global inequalities (as Galtung did) understood the calls for scarcity and a stop to economic growth as an attempt by developed countries to preserve the current status quo and, in so doing, to colonize the future of the global South. For experts from the South, "zero growth" was not an acceptable prospect as this idea was perceived as a serious threat to the Southern chances of industrialization and development.[63] The authors of *The Limits to Growth* had indeed noted that equilibrium politics was to stabilize existing inequalities, for example, in the distribution of food.[64] Nevertheless, the MIT model had not made a differentiation between world regions. Subsequently, a group of Argentine scientists, supported by the Bariloche Foundation, created their own world model.[65]

Despite their sometimes vehemently worded attacks, even most critics of the report recognized that the book could not simply be pushed aside, if only because it had created such a wide response. The report's pithy formulations and its richly metaphorical language, for which Donella H. Meadows was responsible, stuck in the mind, just as the visualizations of the results in multiple illustrations did too. The world model, the exponential growth curves, and the scenarios of equilibrium soon became common knowledge for the educated public. The metaphor of a lily pond that is still half empty on the twenty-ninth day of growth but is already completely overgrown on the thirtieth made the dynamic of exponential growth easily comprehensible:

> A French riddle for children illustrates another aspect of exponential growth— the apparent suddenness with which it approaches a fixed limit. Suppose you own a pond on which a water lily is growing. The lily plant doubles in size each day. If the lily were allowed to grow unchecked, it would completely cover the pond in 30 days, choking off the other forms of life in the water. For a long time the lily plant seems small, and so you decide not to worry about cutting it back until it covers half the pond. On what day will that be? On the twenty-ninth day, of course. You have one day to save your pond.[66]

Subsequently, these metaphors could easily be referenced and recalled in dialogue, and in fact a discursive element was at work here: in a global communication situation, growth and its limits were discussed in all sorts of academic and political circles. Hardly anyone who operated in a related academic field or among political actors could escape this dynamic and had to take a position on the issue—for or against. This was even more the case when, at the end of 1973, the first oil crisis seemed to confirm the hypothesis that resources were finite. Even if in the U.S. government, for example, the prospect of sharply rising oil prices had circulated since the late 1960s, and to this extent one cannot really speak of an "oil price shock," the reduction in oil production by Arab nations was seen as the harbinger of what, in a few decades, would be in store for the whole world.[67] The Swiss daily paper *Landbote* even attributed the "extortionate behavior of the oil sheiks" principally to this idea: "They have drawn their conclusions from the realistic results of futures research, above all from the study of the Club of Rome about the limits to growth."[68] When in 1974 the American Senate arranged a hearing on the subject of key commodities and materials, the theses of *The Limits to Growth* were also discussed.[69]

The scientific and intellectual discussion of the book also conveyed voices critical of culture and consumerism. They demanded a shift away from the "more, more, more" attitude of modern industrial and consumerist society, as did the bestseller *Small is Beautiful* by the German-British social reformer Ernst Fritz Schumacher, which was likewise influenced by the debate surrounding *The Limits to Growth*. A complete renunciation of growth, or "zero growth," appeared admittedly unrealistic to the majority of speakers and in any case seemed impracticable on a global scale. Nevertheless, at least in intellectual discourse, "qualitative growth" was crystallized as a new guiding principle. This enigmatic concept centered on the notion that the previous, economic understanding of growth (as gross domestic product) should be expanded to include environmental and social aspects.[70] The Club of Rome now called too for "another kind of growth," as Alexander King asserted, "It is therefore not growth which is at fault but the present kind of growth and, one might add, our methods of measuring growth."[71] The follow-up to *The Limits to Growth*, a new world model designed by Eduard Pestel together with the Serbian American Mihajlo Mesarović, referred similarly to the goal of "organic growth."[72]

In any case, discussion of the report clearly transcended national boundaries and even overcame the boundaries of the Cold War: despite its Western-centered identity the Club of Rome fostered contact with Jermen Gvishiani, the vice president of the Soviet Committee for Science and Technology. In both camps of the Cold War, scientists and politicians began to address global problems shared by "advanced" industrial societies—for example, environmental pollution.[73] The socialist states appeared officially reserved toward the

Club of Rome due to its bourgeois affiliation, but were nonetheless interested in its model and knowledge about the future of global resources and environment. Thus, the international discourse on *The Limits to Growth* spurred on nascent environmental concerns in the East. Through Gvishiani's mediation the Club of Rome was able to present a follow-up model named "Aid to Policy Tool" in Moscow at the Soviet Academy of Sciences in 1977.[74]

The insights of *The Limits to Growth* were disputed, and the new but amorphous understanding of growth was not capable of winning national or global majorities. Nevertheless, the study had far-reaching effects. It opened up a new scientific field of global modeling: the very criticism of the study led to the design of other world models, commissioned again by the Club of Rome but also by other institutions such as the Argentine Bariloche group. Further, the International Institute for Applied Systems Analysis, created in Laxenburg, Austria, in 1972, embraced global modeling as a central field of study.[75] Scrutinizing the social and political effects of the study, first, large parts of the European Left were opened up to environmental issues. In the United States, a counterculture had already conveyed environmental values and practices into parts of the alternative milieu and the New Left; now, through the debate over growth, the European Left (including, among others, the West German and Dutch Social Democrats) was won over to environmentalist thought bound up with anticapitalist ideas. The environmentalist and antinuclear movement also achieved its decisive impact through the debate over *The Limits to Growth*. In the second place, some conservatives (in part from the field of traditional nature conservation) picked up on questioning economic growth.[76] In this way, conservative and leftist thought partly melted together, and this bond was a central factor in the founding of green parties in Europe—for example, in Britain and West Germany—in the course of the 1970s.[77]

Third, the discussion about growth generally energized Western and international environmental politics, especially since it coincided with the first UN Environment Conference, which took place in Stockholm in the summer of 1972.[78] *The Limits to Growth* stimulated the European Community's environmental policy. West Germany's minister of the interior, Hans-Dietrich Genscher, influenced by the debate over growth, called a conference of European Community environment ministers in the fall of 1972.[79] Likewise, with direct reference to the MIT study, Switzerland embarked on a comprehensive approach to energy politics, as a national variant on exploring ecological limits by the use of long-term environmental planning.[80] At the same time, demands could now be made in the Swiss Parliament for "zero growth" or a slowdown of economic growth.[81]

Finally, the book brought researchers from industrialized and developing countries together in debate, and stimulated international development and population policy. Enforced by Cold War détente, the North–South divide came more sharply into focus in international politics. Calls from countries of the global South for better terms of trade and a "New International Economic Order" were heightened by the scientific and political discussion surrounding *The Limits to Growth*.[82] The debate strengthened new approaches in development policy, for example, the "basic needs" strategy, which concentrated on people's needs rather than on the swift increase of economic growth rates in developing countries. Critical development experts, gathered around the Dutch development economist Jan Tinbergen, produced a study on global development and nutritional problems—commissioned by the Club of Rome, which took on the criticism from the South. Tinbergen's study called for the North to forgo economic growth for the next three decades in order to support the global South. In general, the study gave scientific credibility to ideas of environment and development as interdependent issues of global politics, thus stimulating new conceptions of development and "sustainable development" in the 1980s, which have had a special influence on environmental and development politics right up to the present.[83] Less obviously, as might be guessed from the main thrust of the book, *The Limits to Growth* had effects on the international dialogue about population, which had been infused with neo-Malthusian thought since 1945. Around 1970 one can identify a "global neo-Malthusian moment," in which predictions of overpopulation spread and national and international family-planning programs were created. However, this population politics quickly lost its momentum, precisely because the South demanded to define its own future.[84]

WHAT MADE *THE LIMITS TO GROWTH* A DEFINING TEXT OF ITS ERA?

As Joseph J. Corn states, past eras' visions of the future can be useful for the historical analysis of the time in which these visions arose: "Visions of the future always reflect the experience of the moment as well as memories of the past. They are imaginative constructs that have more to say about the times in which they were made than about the real future, which is, ultimately unknowable."[85] For this reason, it is worth pursuing the question of why a single book produced such a huge reaction and attracted such intense discussion over a long period of time. On this topic, scholars have repeatedly pointed to the clever marketing of the Club of Rome.[86] The almost simultaneous publication in multiple languages, the packaging of the message to appeal to the media and general public, and the political and economic networks the club was able to use certainly contributed to the considerable attention that the study drew; so too did the effect of the blood-curdling vision of the apocalypse, which, even as a warning, aimed to use the prospect of an impending doomsday to provoke a reaction.[87] But these factors alone cannot explain the great success of the publication. Packaging with no content, or a horror story with no connection to social reality, would have met the fate of a passing fad, rather than becoming a long-standing standard reference. The suggestion that the Club of Rome kept the debate alive by publicizing follow-up studies is also not really sufficient. Ultimately, the second report attached considerably less attention to the Club of Rome.[88]

The crucial key to the question of the study's short-term as well as its medium- and long-term resonance lies in its close connectivity to central discourses dominating at the time. One can pick out three types of discourse that the report tapped into, which were of crucial importance to the powerful reception of this text.

First, beginning in the mid-1960s, environmental protection took on an entirely new social significance in a brief space of time. This was the result of a changing social understanding of the environment, which became identifiable in the 1960s, at first in the United States. Among the factors at work here were the American counterculture and the student movement. These were shaped by alternative values and understandings of life, by the quest for authenticity and naturalness. They also epitomized a protest opposing the war against nature, as manifested in nuclear weapons' tests and in the use of chemical weapons in the Vietnam War.[89] In addition, in the United States, within consensus liberalism new concepts of welfare and development circulated, which implicitly or explicitly began to relativize the goal of economic growth that had become the benchmark of prosperity and economic success since the 1940s.[90] In an age of prosperity, governments should not only ensure the material comfort of their citizens (and thereby economic growth), but upgrade both their quality of life and the condition of the environment.[91] Warnings amplified in the media also played a role here, for example, Rachel Carson's *Silent Spring* raised awareness about the consequences of chemical pesticides for human health and for nature, and advocated technical approaches to the problem.[92] In effect, environmental politics developed in the United States in the 1960s as an independent field of activity and policy. Under the Lyndon B. Johnson administration and in the early Nixon era, numerous laws were enacted that aimed not only to conserve "nature" and "wilderness" but also to protect the "environment" of people. Environmental protection became a matter for modern governance, understanding politics as a field of planning and projected feasibility.[93] From 1968 onward, environmental protection was internationalized: Nixon initiated a committee for a "third dimension" of NATO, which would also take environmental problems into account. The OECD and many Western governments (such as the Social Liberal coalition in West Germany) took up environmental protection as a new field of policy and in the summer of 1972 the UN Conference on the Human Environment was to be called in Stockholm.[94] In 1971 in Switzerland over 90 percent of voters and all states in the Swiss Confederation approved the adoption of an article to the federal constitution for the protection of the environment, and in parliamentary elections that took place in August that year, environmental protection was "the chart-topper of the issues hit-parade."[95] All over the world, the media took up the subject. Not only environmental catastrophes such as the *Torrey Canyon* oil spill of 1967 delivered up powerful headlines, new communications technology such as television also created the opportunity to visualize environmental problems.[96] Moreover, public events received an unexpected influx of participation, such as the 1970 Earth Day in the United States, and the events that took place in Europe within the framework of the European Conservation Year.[97] The turn to environmental policy was also manifest on a semantic level: in German, the term *Umwelt* (or "environment" in English) established itself within a few months in 1970 as a label for nature that surrounded people, where before it had not been used in a people-nature context. "Umwelt" superseded previous labels such as "Lebensraum" (living space), "natürliche Lebensbedingungen" (natural living conditions), or "Immissionsschutz" (pollution control). The same holds true for other European languages, such as the use of "environnement" in French.[98]

The crux of this environmental discourse, which became increasingly far-reaching in all social contexts, turned out to be the attitude toward economic growth. At first, environmental politics had been dominated by an understanding of environmental protection focused on technical solutions to environmental problems, rather than aimed at reducing economic growth or even rejecting the modern industrial era.[99] Thus, for example, Jakob Bächtold, longtime president of the deeply traditional Swiss League for the Protection of Nature, had already recognized at the beginning of the 1960s that "growing material needs" and "increasing population" were fundamental dangers to "natural living space." However, he wanted to combat these with comprehensive planning and a reduction in the number of foreigners.[100] In contrast, by the beginning of the 1970s growth itself became the subject of debate.

In fact, around 1970 the English-language book market, and then the international one too, became flooded with works on ecological crises and doomsday scenarios. This is the second reason that the Club of Rome's report encountered an already sensitized readership. Between 1968 and 1972, a succession of studies predicted environmental and population crises on a global scale and made the case for a radical shift away from previous understandings of growth. These global crisis scenarios coincided with a new view of the Earth. In 1966 the U.S. economist Kenneth Boulding and the British economist and environmentalist Barbara Ward popularized the richly symbolic metaphor of "Spaceship Earth."[101] In a global sense the image of "Spaceship Earth" pointed to the fragility of the Earth and the finite nature of its resources, while at the same time emphasizing the human capacity to steer these resources to effective use. With the photo of Earth captured by *Apollo 8* astronaut William Anders at the end of 1968 and the moon landing in the summer of 1969, images of planet Earth traveled around the world. Thus the metaphor took on a whole new appeal, and structured the pattern of argument in the dawning environmental age.[102] In fact, the ecological crisis scenarios that became bestsellers at the beginning of the 1970s always concentrated on humankind's prospects for the future on the global level. Alvin Toffler warned of a *Future Shock*, Paul Ehrlich ignited the *Population Bomb*, and Gordon Rattray Taylor wrote the *Doomsday Book*.[103] Both in its subject matter and its approach, Barry Commoner's 1970 publication, *Science and Survival*, was a direct forerunner of the Meadows report.[104] In addition, a twenty-six-point comprehensive "Blueprint for Survival," published by the English magazine *The Ecologist* in January 1972, attracted a great deal of international attention. This referred to the study *The Limits to Growth*, the main conclusions of which were known to the authors of the "Blueprint" in advance.[105] *The Limits to Growth* both built on these apocalypse scenarios and was at the same time the apex of this form of publication.

Third and finally (and from today's perspective this may be surprising), *The Limits to Growth* built on the somehow utopian ideas of planning and controlling the future circulating in the late 1950s and 1960s. The study understood the future as, if not entirely, then at least to a great extent, predictable and plannable. The results of the study, which from today's perspective rest on an incredibly narrow and shaky basis, were bound up with far-reaching political demands only because there was still a widely prevalent scientific, social, and political belief that the future could be predicted by science and planned by politics. The study's use of systems analysis, modeling, and computers made it compatible with conceptions of planning, automation, and control. These conceptions came out of the natural and engineering sciences, cybernetics, and systems analysis as well as the new field of futures research, and spread into the political discourse in the East and West during the 1960s. The basic starting point was that many researchers were indeed confident that the future could be analyzed and controlled through the use of modern approaches such as cybernetics and through new technological tools such as the computer, enabling them to produce relatively reliable knowledge about the future and to make this knowledge available for political planning. This conviction corresponded to a largely positive understanding of science and technology among broad sections of the political elite, who aimed to incorporate scientific expertise and modern technology into the political process in order to be able to make decisions as rationally and objectively as possible. Thus, prediction and planning instruments created by the RAND Corporation (and other think tanks), such as the Planning Programming Budgeting System circulated among West European governments. Thus, for example, the West German Social Democrats developed a vision of a forward-looking politics, which, through the formulation of middle- and long-term goals and the use of modern technologies, would be able to tackle "newly arising structural changes in business and society."[106] In 1964 the Swiss Social Democrat Martel Gerteis saw in cybernetics the "beginnings of a kind of super-science." In the year 2000, Gerteis predicted, governments would probably employ cyberneticists who, before important decisions were made, would simulate the outcomes of different possible variations with the help of a "brain center."[107] When Meadows and his team described their world model, ideas of such "government machines" were not far off: "A dynamic model deals with the same incomplete information available to an intuitive model, but it allows the organization of information from many different sources into a feedback loop structure that can be exactly analyzed. Once all the assumptions are together and written down, they can be exposed to criticism, and the system's response to alternative policies can be tested."[108] The Limits to Growth owed the wide reaction it provoked, its nature as the focal point of debates about the future of society and politics, and finally also its enduring effect to the fact that the study rested on a historical interface of development, epitomizing a historical conjuncture of different discourses. The Limits to Growth united ecological values and interpretive frameworks with aspirations for reform and the creation of a new society, which drew on the previous decades' passionately optimistic attitude toward planning and controlling the future and a sense that anything was achievable.

Soon, however, the revolution devoured its own children. The different areas of meaning that briefly coalesced at the time of The Limits to Growth soon broke apart once again. The report itself contributed greatly toward profoundly shaking the belief in economic growth and technological advancement, which in the 1950s and 1960s had hardly been questioned. This also contributed to a loss of trust in scientific predictions and expertise, and in generally applicable, global formulas. Systems research and modeling did continue to be the subject of great fascination in academia, and played a major role in global climate research. But the high expectations for the universal validity of computer models, which had been awakened by initial studies such as The Limits to Growth, were increasingly called into question in scientific practice, particularly in the difficulties in putting theories into operation in coherent models. Futures research became aware of the limited capabilities of predictions and accordingly it formed its statements much more carefully than had been done in 1972 in The Limits to Growth.[109] Likewise, the environmental movement subsequently linked up "think globally" with "act locally." The movement stayed alive not by invoking the looming apocalypse but by inventing creative, context-specific alternative futures combining global responsibility and local action. The Club of Rome subsequently released new, specified world models but in the 1980s it was no longer able to build on the impact of The Limits to Growth (which was also caused by a conservative backlash in the Reagan-era United States). It was not until the

1990s that its reports, for example, Ernst Ulrich von Weizsäcker's *Factor Four*, met with greater recognition.[110]

This renewed interest may well be connected to the revitalization of global environmental discourse characterized by conceptions of sustainable development and the challenge of climate change. The theses of *The Limits to Growth* found new attention, and the study once more influenced the global debates over the ecological future of the Earth, even into the new millennium.[111] Amazingly, even decades after its publication, estimations as to how the hypotheses of the report should be judged differed widely. So, for example, Graham Turner of the Commonwealth Scientific and Industrial Research Organisation declared in a quantitative analysis in 2008: "30 years of historical data compare favorably with key features of a business-as-usual scenario called the 'standard run' scenario, which results in collapse of the global system midway through the 21st century."[112] By contrast, precisely the opposite view was taken by Bjørn Lomborg and Olivier Robin in a *Foreign Policy* article in 2009:

> The Club of Rome's most dire forecasts have failed to come true. Vital minerals such as gold, silver, copper, tin, zinc, mercury, lead, tungsten, and oil should have been exhausted by now. But they aren't. Due to an exponential increase in population growth, the world should be facing desperate shortages of arable land and rising food prices. Yet food prices have never been lower. And the world's health should have been undermined by an exponential increase in pollution. People today, however, live longer than ever before, and in Western cities, most pollutants are on the decline, driven down by technological advances and environmental legislation.[113]

The different assessments can be traced back to the fact that the significance of scientific and technical innovation and the substitution of raw materials is judged differently. In Lomborg and Rubin, we see Samuelson's critique again: "Tempting though it might be to attribute these faulty predictions to flawed methodology and bad math, their real weakness is the underlying assumption that planet Earth has finite, essential resources (such as oil, water, and grain) for which there are no substitutes."[114] Moreover, the varying assessments stem from a differing estimation of the scale of environmental pollution, namely, with regard to global climate change. In 1972, *The Limits to Growth* stated that should reserves of natural resources be greater than supposed, environmental pollution would become the decisive problem that would set limits to growth.[115] In this way the study implicitly anticipated the development of global environmental discourse of the following decades, whose emphasis shifted in the 1980s and 1990s, influenced by the dominant theme of climate change, from resources toward waste and emissions.

Judging *The Limits to Growth* today, it seems central that the study (independent of the modeling, which, from a modern-day perspective, seems very elementary) is about a "self-destroying prophecy,"[116] a warning, an alarm. Its intention was to sketch out a thoroughly dramatic crisis scenario of the looming collapse, in order to provoke corresponding social and political reactions. To this extent, the intention of the authors and the Club of Rome has panned out, even if global action has fallen short of their own ambitious expectations.[117]

4

THE SUM OF ALL GERMAN FEARS

FOREST DEATH, ENVIRONMENTAL ACTIVISM, AND THE MEDIA IN 1980S GERMANY

FRANK UEKÖTTER AND KENNETH ANDERS

FEW ISSUES HAVE LEFT a more powerful imprint on German environmentalism than the forest death (*Waldsterben*) debate that held the country in its thrall during the 1980s. The specter of widespread deforestation, usually attributed to airborne pollutants, scared people across the political spectrum, helped the Green Party to enter parliament in 1983, and brought a new dynamism to a heretofore stagnant environmental policy. It effectively drove Western Germany to become the only Western country with an environmental boom during the early 1980s. More than any other issue, it transformed German environmentalism into a mainstream endeavor.

But at the same time, forest death is an invitation to cultural stereotyping. Woodlands have defined the collective imagination of Germany ever since three Roman legions perished in the Battle of the Teutoburg Forest in 9 AD. That made it tempting to depict the forest death scare as the latest incarnation of a national trope. "More than any other nation, Germany has embraced the process of civilization as forest history," Bartholomäus Grill wrote in the popular weekly *Die Zeit* in 1987.[1] In 2011, an exhibition at Berlin's German Historical Museum presented German sylvan mythologies in their full splendor.[2] However, stereotypes of this kind make for difficult intellectual terrain. Even if we leave the dubious assumption of national character aside, these stereo-types are notoriously unspecific: they do not explain why the forest death debate arose at a certain point in time, and they certainly do not prejudge the outcome. It is more rewarding to turn this question around and inquire about the view of forests as it emerged from the debate. How did people perceive woodlands, what were the defining features of the forest under threats, and which features did they ignore? Only after discussing these questions can we make an informed judgment about whether, and in what way, forest death was a German peculiarity.

People had known for more than a century that forests incur harm from air pollution. The first section describes how pollution issues were kept under control, if only in political terms, by all sorts of deals behind the scenes since the late nineteenth century. The second section describes the gradual erosion of this consensus in the postwar years. The discussion's focus moved from metal smelters and other factories with high sulfur dioxide output to coal-fueled power plants. Abatement technology received a boost, and sulfur scrubbers were operational by the early 1970s. At the same time, a new gener-ation of scientists sought to feed their findings into the political discourse. All this was setting the scene for a debate, but when and how this debate would take place was underdetermined.

Two factors transformed the potential debate into a real one. One was press coverage: the notion of forest death was a media cliché par excellence, and forest death marked the emergence of journalism as a force in its own right in environmental debates. The other factor was the economic crisis of the 1980s and the political vacuum that it produced. Unlike other countries, Western Germany lacked a powerful push for the realignment of political coordinates. The economic crisis in the wake of the 1979/80 oil crisis hit Ger-many no less hard than its neighbors, but the widespread feelings of crisis had nowhere else to go. Forest death was the issue that filled this vacuum.

After a discussion of these causal factors, the chapter proceeds with a discussion of short- and long-term consequences. While environmental initiatives played a marginal role in the run-up to the forest death debate, they profited immensely from the fallout. Pollution became a defining issue of green mainstream sentiments, and many environmental groups received a boost—though, interestingly, no new nongovernmental organizations (NGOs) grew out of the forest death debate. The trope also opened a window of opportunity for environmental policy. However, this window closed after only a few years, and forest monitoring, inaugurated under the impression of forest death horror scenarios, turned into a ritual that gradually lost all political potential. The conclusion offers some thoughts on what it means to think about forests in the wake of its presumed demise.

Sulfur emissions knew no boundaries, but the discourse over their consequences did. This chapter focuses on the Federal Republic of Germany (FRG; West Germany) as the place where forest death was invented and where the discourse had its greatest influence. This does not mean that the debate left other countries untouched. While the German Democratic Republic (GDR; East Germany) tried its best to curtail the discussion within the socialist sphere, the effect of West German media reports was so strong that the word even came up in petitions to GDR government agencies.[3] The debate also had a European dimension, as the heightened concern over air pollution played a major role in the development of a common environmental policy within the European Community during the 1980s.[4] However, these international repercussions were a reflection of a vigorous debate within the FRG, and we need to understand the causes and consequences of this national debate before we can discuss effects beyond borders.

A WELL-KNOWN PROBLEM

The term "forest death" entered the German vocabulary in the early 1980s as the standard expression for an imminent sylvan disaster. However, the problem was anything but new: people had known about the harmful effects of sulfur dioxide (SO_2) on plants for more than a century.[5] It was around 1850 that Adolph Stöckhardt, professor of agricultural chemistry at the Saxon Forest Academy in Tharandt, began the first scientific studies of the problem. We can speak of an established tradition of research since 1883 when Julius von Schroeder and Carl Reuss published a voluminous book that looked into smoke damage with particular attention to the Harz Mountains.[6] Complaints from agricultural and forest circles were driving research from the outset, making smoke studies a scientific as much as a political and administrative endeavor. We can even find references to Schroeder and Stöckhardt in a speech in the Prussian House of Representatives.[7] Scientific interest seemed particularly strong compared to other environmental problems, and some researchers found that sulfur dioxide had inspired "the greatest body of literature in the field of air pollution."[8] German smoke research even won acclaim from scientists overseas. As late as 1949, Robert E. Swain, a professor at Stanford University, declared that German smoke studies from the early 1900s continued to deserve "a place in the reference file of every worker in this field."[9]

However, decades of scientific scrutiny did not lead to a consensus on effects and symptoms, and air pollution handbooks declared sternly that researchers were still far away from a comprehensive understanding of the matter.[10] Scientific knowledge remained patchy until far into the twentieth century, but that was not much of an obstacle in the search for solutions. Researchers disagreed about the extent of damage and whether the damage came from exposed leaves or acidic soils, but they were unanimous on the fundamentals: sulfur dioxide could damage vegetation. As a result, companies with high sulfur dioxide emissions were not in a position to deny harmful effects as a matter of principle, and that made them open to all sorts of deals when it came to damage claims. As long as costs remained within certain limits, deals were simply the path of least resistance. It was a matter of credibility, as industrial corporations could not easily ignore scientific evidence. It was also a matter of good community relations, as metal smelters were often located in rural regions, where industrialists could ill afford arrogant behavior. Some—particularly companies that were owned by the state—even felt that fair deals for neighbors were a matter of honor.

As a result, state-owned enterprises were usually the most generous when it came to compensation claims. The state of Saxony, which comprised an old mining region around the town of Freiberg, created a special position in order to keep conflicts with neighbors under control. Farmers and forest owners were supposed to send their complaints to a "smoke commissioner" (*Hütten-rauch-Kommissar*), who routinely investigated the matter and paid out compensation according to findings.[11] In the Siegerland, another region with a long mining tradition, the smelting companies Wissen and Alte Hütte signed ten-year contracts that obliged them to pay 1,658 reichsmarks per annum.[12] In nearby Niedermarsberg, the Kupferbergbau Stadtberge gave two local peasants authority to judge damage claims.[13] A company in Bergisch-Gladbach found yet another solution in purchasing land with pollution damage. It also made a deal with a forest owner to buy wood from his property at a favorable price.[14] When it came to making peace with neighbors, polluting companies were open to all sorts of creative solutions.

To be sure, these solutions were usually the result of tense negotiations. A trade journal wrote that many factory owners saw smoke damage as "a ghost that never rests," and that captured the sentiment nicely: making a deal on pollution damage was akin to a poker game where both sides sought to maximize their returns.[15] But at the end of the day, negotiations were a routine part of doing business, and smoke was just one more issue that industrialists had to keep in mind. In any case, we find a more or less settled routine for damage claims in many German mining regions by 1900. The sums claimed and paid changed over the years, and so did methods and procedures, but the general approach remained the same: it was some kind of compensation that would make the peace. If a company emitted significant amounts of sulfur dioxide, these deals were essentially business as usual. The Reichsanstalt für Wasser- und Luftgüte, the supreme scientific authority on pollution in Germany, reported, "We have cases where these rents have assumed a status resembling customary law."[16]

Of course, some industrialists were more reluctant than others, and some neighbors found that a lawsuit was their last resort—a risky and time-consuming endeavor that even government officials called "a gamble."[17] However, many claimants had a realistic chance to obtain at least a limited payment or some other compensation, and money was usually all that farmers and forest owners cared for. Administrative files rarely include sentimental comments about the devastation that pollution wrought, and the same held true for the call for more comprehensive change. If claimants voiced any ideas about reform at all, they usually concerned making it easier to gain compensation. At a conference of the Saxon Forest Society (Sächsischer Forstverein), one speaker suggested that "it would be good to introduce a tax on coal or acid emissions, with the revenue going to forest owners as compensation for smoke damage that they incurred"—a call for environmental taxation, technically speaking, though the speaker was surely thinking more along the lines of a subsidy.[18] The proposal went nowhere, but it shed a revealing light on the priorities in the forest community. They had their eyes on money and little else.

As a result, reducing sulfur emissions was a second-rate issue into the postwar years. None of the parties found that it was better to tackle the problem at the root rather than seek solutions for the ensuing damage. The government was not behind pollution control either, as officials were usually content when negotiations forestalled an escalation of conflicts. Both sides tried to maximize their gains within the framework of the compensation regime, and that brought the debate to focus on places where sulfur dioxide concentrations were particularly high. As a result, vegetation damage was an issue in the vicinity of metal smelters, chemical factories, and other plants that produced acidic emissions, but it was not an issue for society in general. In other words, it was a problem with limited geographic scope. In 1907, an expert declared at the International Agricultural Convention (*Internationaler Landwirtschaftlicher Kongress*) in Vienna that within all of Prussia, only about 50,000 hectares showed pollution damage to vegetation, with total devastation existing on only about 5,000 hectares.[19]

While metal smelters and other corporations were spending a lot of time haggling with their neighbors, there was little talk about sulfur dioxide from the combustion of coal. However, these emissions were no small matter. Schroeder and Reuss calculated that Germany had used about 33 million tons of coal in 1872 and that their combustion set off about 800,000 tons of sulfur dioxide.[20] Wherever coal was burned in great quantities, the atmosphere showed evidence of contamination. In 1911, a municipal research bureau in Leipzig left litmus paper in the ambient air of Leipzig overnight. Litmus paper is commonly used in chemistry as a pH indicator, and officials found the next morning that the paper had turned red, indicating acidity.[21] "As cities are expanding and are home to many factories, there are many instances of damage to vegetation from emissions," an agricultural journal noted in 1923.[22] The following year, the handbook for plant diseases (*Handbuch der Pflanzenkrankheiten*) noted, somewhat pathetically, that "with every smokestack, we are building a plant poisoning device."[23] But for the time being, a struggle against this type of pollution was beyond debate. "As it stands, nobody has given serious thought to the absorption of acidic emissions from coal combustion, and all that remains for the moment is to let these gases escape into the open air," Schroeder and Reuss wrote in 1883.[24] It was a stance that would remain unchallenged for decades.

To some extent, apathy stemmed from a lack of abatement technology. Hans Wislicenus, one of the leading authorities on smelter smoke, wrote on the eve of the First World War that sulfur dioxide concentrations in the exhaust gases of contemporary scrubbers were actually higher than the concentrations in the unfiltered exhausts of coal furnaces.[25] A German company, the Metallgesellschaft, came forward with new sulfur scrubbers in the 1930s that worked with a minimum sulfur dioxide concentration of 1.5 percent, prompting industrial leaders to proclaim "great progress."[26] However, contemporary experts gave 0.08 percent as a typical sulfur dioxide concentration in coal exhausts.[27] Of course, that figure was open to improvement, but given the prevailing focus on monetary compensation, no one really pursued research on coal exhausts cleaning devices with vigor. Some authors called for a comprehensive "war on smoke," but in the absence of a lobby for stack gas cleaning, that was just empty rhetoric.[28]

In sum, the sulfur damage to plants was a problem under control, though only in political terms. All parties agreed to treat it as a local problem, devoid of broader significance. Control could take many forms, and abatement was not inherently superior to compensation: both were legitimate means of conflict resolution, and the choice was made in the light of the local specifics. The Steinkohlen-Elektrizitäts-AG of Essen, a major operator of power plants, declared as late as 1961 that "it is often cheaper for corporations to provide payments for some small-scale damage than to invest significantly larger sums for prevention."[29] As things stood, acidic pollution was simply something between polluters and their neighbors. But that changed in the postwar years.

A CONSENSUS ERODING

The consensus showed signs of fracturing in the early postwar years. Doubts emerged from several sides. Calls for abatement became stronger, and monetary compensation appeared increasingly dubious. Research on technological fixes received a boost, finally making desulfurization of coal-fired power plants a realistic option. As smokestacks rose to unprecedented heights, it became elusive to attribute forest damage to a specific polluter. Monitoring showed evidence of long-range damage to Scandinavian lakes. A new generation of scientists felt comfortable to represent their findings in the political arena and provided new vigor to the calls for abatement. Yet these trends were much more effective in undermining the nineteenth-century consensus than in pointing the way to the future. By the 1970s, the age of shoddy deals behind the scenes was over, but it was not clear what the replacement would be.

Discussions over new environmental policies began in the 1950s, with the state of North Rhine-Westphalia in the vanguard. The original impulse came from problems in the heavily industrialized Ruhr region, where people were no longer willing to tolerate the excessive pollution that went along with coal mining and steel production. Popular opinion prioritized visible problems such as smoke and dust, but once those protests had inspired a comprehensive look at air pollution problems, sulfur emissions were on the agenda.[30] "It's not dust that is the most dangerous—it is sulfur dioxide," an official from the Federal Ministry of Labor (Bundesarbeitsministerium) declared in 1958.[31] Four years later, Heinrich Lent, the powerful chairman of the Clean Air Commission of the Society of German Engineers (Kommission "Reinhaltung der Luft" beim Verein deutscher Ingenieure), stated at a conference, "All participants know that we need to solve the sulfur dioxide problem with reasonable technological means and at reasonable costs over the upcoming decades."[32]

Cleaning devices were still lacking, but since the 1960s, it was considered only a matter of time before flue gas desulfurization technology would be developed. In 1964, the new German manual for air pollution control technology (*Technische Anleitung zur Reinhaltung der Luft*) included a clause requiring that when licensing new boilers, authorities should keep the future development of sulfur dioxide scrubbers in mind: once sulfur dioxide emissions were above a certain threshold, authorities should "check whether the operator needs to leave some space" for these devices.[33] In North Rhine-Westphalia, the minister of commerce and industry took the view that leaving space "was a must for plants with a high sulfur dioxide output."[34] Werner Figgen, the minister in charge of air pollution control in North Rhine-Westphalia, declared in May 1968 that flue gas desulfurization was "a number one priority" and expressed confidence that "we will have the technology to solve the problem a few years from now."[35] His ministry sponsored research and development at one of the state's industrial conglomerates, the Grillo-Werke, and that company was quick to proclaim success. A journal article of 1971 declared that the company could "build and operate flue gas desulfurization plants of any size."[36]

Of course, press releases were not a sound base for political decisions. But independent experts soon chimed in: new technology made it possible to remove sulfur dioxide from exhausts. After study trips abroad, the Associations for Technological Inspections (*Technologische Überwachungs-Vereine*), whose independent expert statements were in high regard beyond Germany, declared in 1974 that flue gas desulfurization technology was used successfully in Japan and thus "state of the art" (*Stand der Technik*).[37] Nonetheless, it was three more years before a power plant in Wilhelmshaven became the first in Germany to use such a device to reduce its sulfur dioxide emissions.[38]

Cleanup continued to advance at a glacial pace over the following years. Of the ninety power plants in operation in Western Germany by the end of 1982, only seven had flue gas desulfurization technology.[39]

Removing sulfur dioxide from coal exhausts was expensive, and industry sought to delay action from the beginning. It required long and tense negotiations to agree on the first ambient air quality standards for sulfur dioxide in 1961, and the vested interests did not relent over the following years.[40] The makers and operators of power plants were a powerful group in Germany, and they did not mince words. In 1974, the Deutsche Babcock AG wrote a letter that lambasted "unqualified and hysterical environmental standards" and declared that three thousand jobs were at risk.[41] Trade union officials were no less concerned and spoke about "nonsense," "kamikaze," and "self-destructive hysteria" when the Federal Ministry of the Interior presented a policy draft for the control of sulfur dioxide emissions from coal-fired power plants: if implemented, the draft would unleash "economic chaos."[42] Industry did not even rethink its stance in the face of generous government proposals. In the late 1970s, the state of North Rhine-Westphalia offered to pay 50 percent of investment costs for new flue gas desulfurization plants, only to find that no company wanted to make such a deal.[43]

However, obstruction from industry was not the only reason for the stagnation of environmental policy. Farmers and forest owners maintained their focus on compensation, and that drained much of the vigor from the call for abatement. It was a somewhat odd stance in a society that was increasingly concerned about the effects of air pollution, and yet foresters stuck to their traditional preferences. "The German law is completely deficient when it comes to dealing with compensation claims," a petition from forest owners in North Rhine-Westphalia stated in 1954.[44] A few years later, the Federal Ministry of Agriculture even opposed a bill that sought to strengthen air pollution, as it felt that the bill would diminish the prospects of neighbors in their quest for compensation.[45] In 1971, the ministry of agriculture sought to insert a clause into a new federal law against air pollution that would have given neighbors the right to turn to every polluter and claim payments without an obligation to consider the polluter's contribution in relation to total emissions—a clause that would have invited abuse. Officials from other ministries found the idea so odd that they did not even include it in the meeting minutes.[46]

This stance left the control of sulfur dioxide emissions without a powerful lobby. Why should officials urge the operators of power plants to install scrubbers when their neighbors were more interested in the prospects of damage claims? But then, the stance of forestry had its own rationale, and it was not just about making money from forests before maturity. Forestry was a distinct social sphere: it had its own set of institutions, special schools, particular career paths, and a long and proud tradition that ultimately went back to the wood scarcity scare.[47] All this created a notable distance from the rest of society, and foresters felt under no obligation to join the broad debate over environmental issues in postwar German society. The shock that the forest death scare implied for the forest community was also about the end of a splendid isolation.

While power plant owners were skeptical of sulfur scrubbers, they had more sympathy for another technological device: tall smokestacks. Environmentalists were aghast, and their critique could draw on a long tradition.[48] For example, Hans Wislicenus once called tall smokestacks "giant artillery guns for long-distance attacks against large forests."[49] They were stopgap measures at best, poised to "achieve a wide distribution of pollutants" instead of a real solution.[50] A Bavarian ministry called distribution a "surrogate solution" (*Ersatzmaßnahme*) in 1969.[51] Experts warned that dilution was also dubious in view of growing doubts about existing ambient air quality standards. For example, Wilhelm Knabe, a government expert who would later become a prominent figure in the Green Party, wrote in an article of 1972 that "we can expect damages far below the threshold levels for sulfur dioxide as they currently exist in the Federal Republic of Germany."[52] As power plant owners were building tall smokestacks in the 1960s and 1970s, they were building living anachronisms.

The German debate was effectively stalled. Things looked more dynamic in an international context. While sulfur dioxide had traditionally been a local topic, transnational pollution emerged as a political issue in the postwar years. Researchers showed as early as 1957 that a significant amount of sulfur dioxide traveled across national borders.[53] By the end of the 1960s, the consequences began to show, as pH values in Scandinavian lakes revealed growing acidification.[54] Sweden raised the issue within the Organization for Economic Cooperation and Development (OECD) in 1969, and the OECD began a research and monitoring program on transnational air pollution in 1973.[55] In 1976, Sweden started a nationwide program to neutralize unwanted acids in lakes and rivers through liming.[56] Sweden became a trailblazer for transnational air pollution control, and yet it seems that they made a strategic mistake when they raised the issue of compensation. It was a game changer, as it had been in previous decades: talk about abatement ceased, and the parties went into poker mode. According to Edda Müller, an independent-minded government official who wrote a book on West German environmental policy from the inside, the government made a cabinet-level decision that effectively declared long-range transports of sulfur dioxide a nonissue so as to avoid playing into the hands of Sweden.[57] Once more, it showed that the discourse on air pollution could focus on either compensation or abatement, but not on both.

On November 11, 1977, the German cabinet voted to authorize work on a Large Furnace Ordinance (Großfeuerungsanlagen-Verordnung), but that quickly turned into yet one more sign of policy stagnation.[58] The first draft surfaced in May 1978, but opposition from industry prompted the government to delay action.[59] Things looked more dynamic on an international level. In 1976, the European Commission began negotiations over a directive on ambient air quality standards for sulfur dioxide and dust in urban agglomerations.[60] Three years later, Germany signed the Geneva Convention on Long-range Transboundary Air Pollution.[61] In other words, it was an open question whether solutions to the acid rain problem would be national or international in nature. With the German government essentially kicking the can down the road, it looked as if sulfur dioxide might become a key issue for an emerging transnational environmental polity.

Among the many issues that plagued German policymakers, perhaps the thorniest was retrofitting. Power plants were long-term investments, and retrofitting was traditionally one of the touchiest issues of air pollution control: it disrupted operations, brought additional costs, and generally nourished a sense of uncertainty that managers abhorred. Concerns over deficient implementation were a running theme in German environmental discussions ever since a landmark report of the Expert Council for Environmental Questions (Sachverständigenrat für Umweltfragen) had raised the issue in 1974.[62] German officials were traditionally inclined to delay action and water down standards under pressure, and the licensing procedures of power plants provided showcases for their leniency. The most glaring case in point was Buschhaus, a 350 megawatt power plant near Helmstedt close to the GDR border. The original plans projected annual sulfur dioxide emissions in the range of 150,000 tons, a whopping 6 percent of Germany's total power plant emissions, while providing only 0.4 percent of electric power.[63] In 1981, a watchdog report from the Berlin Social Science Center (Wissenschaftszentrum Berlin) expected a significant rise in German sulfur dioxide emissions for the 1980s.[64]

Changes were also under way in the research community. Dealing with sulfur dioxide emissions was a science-based endeavor from the outset, but researchers typically felt more comfortable at a distance from politics. Their favorite place was the lab, where they could study plant samples or expose trees to sulfur emission in special greenhouses. Schroeder and Reuss even distinguished between a government view that was interested in clear figures and a scientific view that stressed "the uncertainty and diversity of observations in different places" and thus abstained from generalizations.[65] However, key experts of the forest death debate such as Peter Schütt and Bernd Ulrich were part of a new generation of forest researchers who combined academic and political thinking—probably a result of the shakeup of German academia in the wake of 1968. They came out of basic forestry research, which put them at a distance from both the smoke research tradition and the clientelist tradition that focused on the financial interests of forest owners. Their interest lay in a healthy forest, and that brought them to see pollution as a serious threat.

Schütt was a botanist who studied Southern German fir trees. These trees showed growing signs of trouble in the 1970s, but Schütt was unable to identify a clear cause. The term "fir death" (Tannensterben) came into use in expert circles, preceding "forest death" in more than a terminological respect. Earlier generations of researchers would have called for more research, but Schütt did not stop at writing grant applications. He aimed for an audience beyond academia.[66] Schütt and Ulrich wanted to alert the general public, and their statements showed remarkable skills in catering to the needs of politicians and the media.[67] Ulrich obviously talked about more than scientific findings when he called the problems of the forests "writings on the wall."[68]

All in all, the consensus of earlier decades was clearly eroding, but it is unclear what would take its place. A debate was in the air, but its trajectory and outcome were highly unclear. Would it focus on tall smokestacks, or Swedish lakes, or deficient standards, or money-hungry foresters? Or would authorities continue to kick the can down the road? Western Germany had been dragging its feet for years, and it was under no immediate pressure to change this attitude. Schütt and Ulrich were making their points, but they clearly needed a receptive audience in order to be heard. As it turned out, that audience was not where one might expect it.

FIGURE 4.1. "Acid Rain Over Germany—The Forest is Dying." This cover story in *Der Spiegel* marked the breakthrough of forest death in the West German public. It was a product of a new generation of journalists: more than just covering environmental issues with a measure of sympathy, as had been the case in the 1970s, they outlined dramatic scenarios that exceeded even the rhetoric in environmental circles.

A MEDIA MYTH

In theory, the environmental movement would have been the natural defender of the dying woodlands. West German environmentalism had gathered significant strength during the 1970s, culminating in the formation of the Green Party toward the end of the decade.[69] But in practice, the burgeoning movement was initially reluctant to put its weight behind the cause.

Nuclear power was the defining issue of German environmentalism since the mid-1970s, and that caused activists to look at coal-fired power plants as a lesser evil.[70] In the coal state of North Rhine-Westphalia, the leading environmental NGO, the Bund Natur- und Umweltschutz Nordrhein-Westfalen, was an ardent supporter of the government's *Kohlevorrangpolitik*, a preferential policy for domestic coal.[71] Even environmentalists who were concerned about the future of Germany's woodlands had their eyes on problems other than pollution. Horst Stern published a voluminous book titled *Save the Forests* (*Rettet den Wald*) in 1979, which focused only superficially on air pollution. When it came to the extent of the problem, Stern cited the old figure of fifty thousand hectares of damaged forests, which was less than 1 percent of Germany's woodland total.[72]

Things looked different in the media. Journalists had covered environmental issues throughout the 1970s, but they were generally transmitters of sentiments rather than protagonists. This began to change toward the end of the decade. The book *Seveso Is Everywhere* (*Seveso ist überall*) of 1978 was a case in point: it was a coproduction of Fritz Vahrenholt of the Federal Environment Agency (Umweltbundesamt) and investigative journalist Egmont R. Koch.[73] The forest death discourse provided further evidence of a more assertive role of journalists, as media coverage presented the issue in a way that maximized popular concern. A three-part series in the muckraking German weekly *Der Spiegel* provided a showcase for this framing in November 1981, when the debate was about to take off. The magazine combined dramatic statements from scientists ("a ticking time bomb," "the greatest environmental disaster ever") with references to lung cancer and Cold War fears: it declared that tall smokestacks were "comparable in range with nuclear missiles" and compared the effect of emissions with chemical weapons—the only consolation being that they did not work "quite so fast."[74]

Aggressive media coverage rather than environmental activism pushed forest death onto the political agenda.[75] Media scholar Rudi Holzberger has sketched the trajectory of the discussion as follows. The state of the forests received an increasing share of attention in 1981 and 1982, then peaked in 1983 and stayed at a high level until 1986, when a gradual and discontinuous decline set in.[76] However, Holzberger was less than impressed with the sophistication of media reporting. He found many articles clichéd and strangely one-dimensional; nobody would have guessed from these reports that forests were the most complicated terrestrial ecosystem. Journalists liked to dwell on generalities and spoke about "the death" of "the forest." Most of them did not bother to see the purportedly dying forests by themselves. In Holzberger's sample of ninety-two articles, only four showed evidence of firsthand experience.[77]

We can follow the trail of clichés down to the term "forest death." The rationale behind the term was essentially anthropomorphic, as it imagined a diseased forest on a more or less straight path to its ultimate collapse. However, decades of smoke research had shown that forests could survive high ambient concentrations of sulfur dioxide for many years: trees would show symptoms of all sorts of troubles, but they would not inevitably die. Smoke experts of the early 1900s commonly distinguished between acute and chronic smoke damage. "In the case of chronic smoke damage, plants do not perish even over long periods of time, but they do suffer from a number of constraints on development that result in reduced growth or less bountiful harvests in the case of agricultural crops."[78] The choice of words was revealing: smoke damage was a problem, but not a looming disaster. The forest death debate favored a different rationale: it perceived stunted growth as a lingering illness that called for strong action. A problem that industrial society had been living with for many years suddenly became a matter of life and death.

CRISIS IN THE AIR

The forest death debate coincided with a time that was ripe with crisis scenarios. Historians increasingly view the early 1980s as a watershed of modern history. The second oil price shock rocked Western economies and brought a repeat of the stagflation crisis of 1973/74. Mass unemployment was straining the resources of Western welfare states. The Cold War came to another climax with the Soviet invasion of Afghanistan, martial law in Poland, and NATO rearmament. Neoliberal governments in the United States and Great Britain brought about a permanent realignment of the lines of political conflict.

Due to its dependence on exports, the West German economy suffered severely from the global economic downturn. Mass unemployment also evoked traumatic memories of the Great Depression that had allowed the Nazis to seize power. But unlike the United States, Great Britain, and France (where Mitterrand banked on nationalization after winning the presidency in 1981), Western Germany lacked a new political project. The governing coalition of social democrats and liberals, in office since 1969, had won another federal election in 1980. However, the economic crisis split the coalition in two camps, resulting in a prolonged process of mutual disaffection that ended with the coalition's collapse in September 1982. The liberal party formed a new government with the Christian Democrats and brought Helmut Kohl to power. But even after this shift of allegiances, the political realignment was modest by international standards. Grand promises of a "moral turn" (*geistig-moralische Wende*) gave way to politics as usual.[79]

Forest death emerged as a political issue during these years, and it is crucial to see it against the background of this malaise. The forest death discourse was one of the most significant crisis discourses in Germany during these years, and it carried the hallmarks of an evasive action from a socioeconomic crisis that defied familiar responses. It provided people with a quantum of solace that, for all the troubles in the economy, there was also an environmental crisis that was probably even deeper. Furthermore, there were some solutions at hand for the forest death problem, rather unlike mass unemployment, which left politicians clueless.

It all came down to a peculiar German response to the crisis of the early 1980s: where other countries immersed themselves in bitter disputes over neoliberal policies, Germany went green. Forest death provided a path to cope with crisis sentiments in a way that avoided more comprehensive political or socioeconomic realignments. The environment was an alternative playing field that served and mirrored the political instincts of West German society, most prominently its commitment to the political middle ground. After the traumatic experience with Nazi extremism, centrism was so deeply ingrained in West German society that some observers have touted it as a proverbial *Volk der Mitte*. Characteristically, the response to the prospect of dying forests cut across the familiar lines of political conflict and stressed a collective interest in forestalling the looming disaster. In a society being torn apart by the centrifugal forces of late twentieth-century Western societies, the environment was one of the last issues that united most Germans.[80]

MAINSTREAMING ENVIRONMENTALISM

For once, Hans-Jochen Vogel was prophetic. Vogel was a dry figure even by the standards of a body politic that was deeply distrustful of charisma after Hitler. But Vogel sensed that forest death would be big: "When the trees are at stake, we will get a people's movement."[81] The environmental historian Jens Ivo Engels later confirmed his assessment, arguing that forest death was "the decisive factor for the 'normalization' of environmental protection in Western Germany."[82] Even Franz Josef Strauß, the stalwart of German conservatism and a veritable hate figure among the German Left, felt that the issue called for a suspension of the culture wars that he usually stoked on every occasion. "We love our forests, forests are part of our homeland" (*Heimat*), Strauß declared, adding that it was time to "do whatever is possible to avert this threat."[83]

The crisis brought environmentalists into the spotlight as never before, and with the severity of the crisis beyond debate, they used the opportunity to make broader claims about humans and the natural environment. Forest death became the symbol of environmental crisis par excellence. "Forest death is a harbinger of a looming disaster that will be even more catastrophic," the Green Party warned in May 1983.[84] "There is no way to deny that we are ruining the forest with our modern lifestyle," wrote Hoimar von Ditfurth, one of Germany's leading environmental intellectuals. Another green spirit, Carl Amery, pointed to the unfolding disaster in a call for "a total restructuring of our systems of production and reproduction."[85] Environmental activists had a field day. They climbed the smokestack of the coal-fired power plant Wedel near Hamburg and threw badly damaged conifers over the Berlin Wall as a reminder that pollution did not stop at the Iron Curtain. In the Black Forest, activists put a huge campaign banner over some 1,500 square meters of woodland, sheepishly telling journalists that they sought to protect the forest "at least for a day" from acid rain.[86] It paid off in attention and more. Environmentalists were riding a wave of concern and raised their presence in Germany's political system to a new level. Thanks to forest death, the Green Party passed the 5 percent threshold in the federal election of March 1983 and thus entered Parliament for the first time with 5.6 percent of the vote.[87]

But for all the vigor of forest death sentiments, it is striking that the debate did not lead to the formation of new civic organizations. No "guardians of the forests" grew out of the debate: the Green Party and environmental NGOs managed to adjust their agendas in ways that suited the needs of the day. In May 1984, the environmental community decided to pool its activities in a special Action Committee against Forest Death (Deutsche Aktionsgemeinschaft gegen das Waldsterben), which made a point of including the more traditional branches of the German conservation movement like alpinism, hunting, sport fishing, and the friends of the homeland (*Heimat*). But in the end, the Action Committee really was not all that much about action. After only a year, it fell "into a long deep sleep" (*Dornröschenschlaf*).[88]

The demise provides a fitting reflection of the trajectory of the forest death debate. The Action Committee was mostly about breathing new life into a debate that was already beyond its prime in May 1984. Environmentalists wanted to extend the green hour as much as possible, all the more so because the debate was taking place in the wake of the deeply divisive debate over nuclear power.[89] Whereas fears about atomic power plants had split the German population into proponents and opponents, forest death brought it back together, and it is no overstatement to call the debate a collective effort at healing. Antinuclear protest culminated in exceedingly violent demonstrations at construction sites, and many people found this deeply disturbing. They gratefully embraced a more consensual issue.

As it turned out, it was not much of a change for nuclear activists. Coal-fired power plants belonged to the same companies that built and maintained nuclear reactors, and these companies were responding to the burgeoning debate with the same kind of arrogance. The polluters were also a perfect enemy for a new Left that thrived on anticapitalist sentiments.[90] And activists could take to the streets and show their concerns in the way that the post-1968 tradition demanded. In November 1984, environmentalists organized a large demonstration in the heart of Munich with speakers from the German Alpine Society (Deutscher Alpenverein) and the Bavarian Trade Union Council (Deutscher Gewerkschaftsbund, Landesverband Bayern). Music came from the quintessential Bavarian antiestablishment band, the Biermösl-Blos'n.[91]

The cumulative result was that environmental sentiments entered the political mainstream, and forest death opened the door for a green decade that stands out as unique among Western countries. In fact, the mainstreaming of environmentalism became an unexpected problem for activists, as they were struggling to find new things to say. How do you flag a problem that everyone is concerned about? One way was rhetorical escalation, which brought environmentalists into a veritable arms race with journalists for the most dramatic scenario. A press release from the Action Committee against Forest Death noted that "hundreds of thousands of jobs were at risk," that "Germany's supply of drinking water was in imminent danger" and that "entire regions" were about to become "uninhabitable."[92] Only days after the landmark report in *Der Spiegel*, the German League for Environment and Nature Protection (Bund für Umwelt und Naturschutz Deutschland, or BUND) put out a press release that warned of "environmental hara-kiri," alluding to the Japanese suicide ritual.[93] The BUND's Bavarian branch, the Bavarian League for Nature Protection (Bund Naturschutz in Bayern), issued a pamphlet titled, "As the Forests Die, So Dies Our Future."[94]

Some quotations look morally ambiguous in retrospect, particularly when people began to cite the Holocaust by way of comparison.[95] But this was not a time for nuance. Even the editor of *Umweltmagazin*, an otherwise dire journal on environmental technology, alleged that he was "scared of rain" and compared the growing extent of forest damage to the cancerous growth of a tumor.[96] Another way to gain attention, and arguably a more dubious one, was the construction of artificial enemies. A federal official was aghast when he joined a panel discussion of the Bavarian League for Nature Protection in Ansbach and found himself on the spot for being in bed with the fertilizer industry—after all, the government's action program included the use of lime in forests so as to curtail acidification of the soil. Did that not weaken the case for drastic curbs on air pollution? The official became so angry that he staunchly rebuked the allegation first in public and then again in private.[97]

While environmentalism was thriving, the forest community was more on the defensive. Against the background of a century of smoke disputes, it was a remarkable change of the guard: forest owners, traditionally the group that expressed the greatest concerns about acid emissions, were now on the margins. A Bavarian forester even declared that "about 60 percent" of forest owners were "either unwilling or unable to grasp the gravity of the situation."[98] Forest owners and environmentalists had different visions of a proper forest, and they had clashed before; Horst Stern's book, for one, was a staunch criticism of narrow-minded profit-seeking in forestry.[99] The debate centered particularly on coniferous monocultures, and as these looked particularly vulnerable to sulfur emissions, the discussion took on a new urgency in the wake of forest death. "Large monocultures should give way to mixed stands," as an assortment of different trees and age groups would be more resilient to insects and pollution, a research institute of the Fraunhofer Society noted.[100] Foresters responded by positioning themselves as innocent victims. "As forestry has not caused these damages to our woodlands, our means cannot have any major effect against them," the German Federation of German Forest Owners (Arbeitsgemeinschaft deutscher Waldbesitzerverbände) argued, adding that adjustments in forest practice would amount to little more than "euthanasia."[101]

Stocking and stand diversity were not the only issues that had forest owners following their traditional predilections. The struggle over compensation entered another round, but while these claims had traditionally focused on industrial companies, forest owners now had their eyes on the state: with damage to forests presumed universal, they felt that it was the government's duty to come up with a law for compensation payments.[102] It was a high stakes gamble, and the forest owners sought to enlist legal and political support.[103] For example, the powerful German Peasants' Association (Deutscher Bauernverband) threw its weight behind the cause: forest owners "expect that a state whose constitution protects the right to property will support them forcefully in the face of an existential threat."[104] It took a speech of chancellor Helmut Kohl at the convention of the German Federation of German Forest Owners to end the discussion.[105] Kohl's stance was purportedly about the legal obligations of the modern state—he argued that there was "no right to general compensation on the basis of existing law,"[106] but there was more to the government's stance. Compensation claims were essentially anachronistic in a society that sought to save the forests, rather than make money with a corpse. In the end, the only thing that forest owners achieved with their quest for compensation was that the gap between forestry and the rest of society became even wider.

Perhaps the most remarkable feature of the forest death debate is the absence of critical voices. Even industrial organizations were hesitant to proclaim that it was much ado about nothing; at the most, they called the effects of acid rain "a completely open question."[107] Tempers obviously varied, but everyone agreed that there was a serious crisis out in the woods—down to the German Federation of Sawmills (Vereinigung Deutscher Sägewerksverbände), which asked for government help in 1983 because forest death would lead to clear-cutting on a grand scale, which would in turn flood the wood market.[108] In 1985, the German postal service put the consensus on a commemorative stamp that showed dying conifers and a stylized clock a few minutes before midnight along with the slogan "Save the Forests."[109] The postal service, not known for an interest in divisive politics, obviously felt that no one could argue with this.

A WINDOW OF OPPORTUNITY

It was clear that the widespread concern called for a political response, but the nature of this response was largely undetermined. Even the primary culprit was open to debate, as experts were circulating widely different models. Some researchers estimated the number of competing explanations as high as 200.[110] Various people sought to establish a link between dying forests and radioactivity so as to build a bridge to the fight against nuclear power.[111] However, air pollution emerged as the most popular cause, specifically sulfur dioxide and nitrogen oxides. For instance, air pollution stood out front and center in an open letter to the German government that 132 forest researchers signed in March 1983.[112] That left its mark on government policies, and forest death provided a powerful boost to air pollution control.

The long-delayed Large Furnace Ordinance moved to the center of political discussions, and politicians of diverse stripes soon began to upstage each other in the quest for the toughest stance. In May 1982, the draft of the ordinance called for 650 milligrams per cubic meter as the maximum flue gas concentration for sulfur dioxide and a ten-year grace period for retrofitting, with the possibility of fifteen years in exceptional cases.[113] By the fall, the maximum sulfur dioxide concentration was down to 400 milligrams per cubic meter. By way of comparison, a contemporary Austrian decree included threshold concentrations of 1,000 milligrams for lignite and 850 milligrams for all other fuels.[114] Interestingly, 400 milligrams per cubic meter was exactly what one of the leading environmental organizations, the Federation of Citizen Initiatives on the Environment (Bundesverband Bürgerinitiativen Umweltschutz, or BBU), had called for in 1980.[115] However, the BBU took a tougher stance when it realized that more was in the cards and called for technology-forcing standards of 200 milligrams per cubic meter immediately and 100 milligrams from 1985. As existing flue gas desulfurization devices were already operating in the range of the draft's limits, the BBU found the proposal reminiscent of "speed limits for snails."[116] When the Large Furnace Ordinance went into force on July 1, 1983, 400 milligrams per cubic meter was the limit for sulfur dioxide in flue gas, a figure that drew criticism even from the youth division (Junge Union) of the ruling Christian-Democratic Party.[117]

The grace period shrunk to a mere five years, which moved the deadline very close to what was technologically possible without major disruptions.[118] The ordinance resulted in a flurry of activities that had no precedent in the history of German air pollution control. According to a study that the Organization of German Utilities (Vereinigung Deutscher Elektrizitäts-werke) published in the late 1980s, the owners of electric power plants made total investments for sulfur dioxide reduction in the amount of 14.2 billion deutschmarks.[119] The decrease in sulfur emissions was even more dramatic than that hoped for in the government's optimistic expectations. At an OECD meeting on the environment in June 1984, representatives of the Federal Republic proclaimed that they would cut sulfur dioxide emissions in half by 1995.[120] In reality, total emissions decreased by 72 percent from 1980 to 1990. For power plants, the decrease was 84.3 percent.[121]

At first glance, power plants were the focus because they were the biggest emitters of sulfur dioxide. According to figures from the Federal Environment Agency, power plants and heating plants contributed 62.1 percent to Western Germany's total sulfur dioxide emissions in 1982. Industry stood at 25.2 percent, households and small furnaces at 9.3 percent, and traffic at 3.4 percent.[122] However, the case against the utilities was equally about pollutants and politics. It focused popular anger on a single culprit and drew a wedge between the energy companies and the rest of industry. Peter Menke-Glückert, the mastermind of federal environmental policy since the 1970s, outlined this strategy in a memorandum of May 6, 1982. In view of the upcoming "hefty confrontation" with industry in hearings, it defined "the isolation of the power plant sector" as "the defining tactical goal." With a relatively generous treatment of other industrial polluters and the refineries, Menke-Glückert sought to survive the political storm.[123] In other words, the political fallout of the forest death debate was not simply about the power of public opinion. It was due to the clever interaction between environmentalists and policy brokers, which was one of the hallmarks of Germany's environmental boom years.[124]

As it turned out, industry did fight tooth and nail against the Large Furnace Ordinance.[125] As late as May 1983, the German economic minister Otto Graf Lambsdorff made a last-minute pitch for yet another hearing of industry.[126] That obstruction was about more than costs: it was an effort to preserve the cozy old days of insider relations in the utility business. For the energy business, forest death was part of a long adjustment process in which companies learned to cope with divergent opinions and multiple stakeholders in society. Energy politics was no longer about networks of "old boys" but about making a case in a contested political arena. After the demise of the forest death debate, some companies launched public relations campaigns to advertise their alleged commitment to environmental protection—with scant information about the subsidies for cleaning devices and the fact that the utilities could write off the costs easily by raising power rates. Public relations efforts were about the image of energy giants as much as about healing wounds. The forest death debate brought a resounding and unexpected defeat for a powerful and self-confident cluster of corporations, and the claiming of environmental credentials was also an effort to regain the initiative.[127]

The government could also claim a reduction in nitrogen oxide emissions, but progress was slower on this issue. Total emissions fell from 2,617,000 tons in 1980 to 1,962,000 tons in 1990 and 1,766,000 tons in 1994. Power plants cut their nitrogen oxide emissions by 60 percent from 1980 to 1994, but the pollution load from transportation remained stable over the decade and decreased only slowly after 1990.[128] Traffic accounted for more than half of total nitrogen oxide emissions.[129] These figures mirrored a government policy that was notably softer on car manufacturers than on the utilities. The government shelved its original plan to make catalytic converters mandatory by 1986 and opted for a more gradual introduction. In the end, the first new cars with mandatory exhaust cleaning were sold in 1988.[130] Faced with the initial deadline, the automobile lobby came up with widely overblown cost projections and, faced with what was arguably Europe's most powerful industrial complex, the government caved in.[131]

Furthermore, the transnational nature of European car transport called for a multinational approach, and the European Community was dragging its feet. It took until March 1985 to find a consensus, and it was not the finest hour of European environmental policy; an expert from the Federal Environment Agency spoke of a "weak EC-compromise."[132] To the best of its abilities, the West German government had pushed for a more vigorous solution, but its role as the vanguard of environmentalism was ultimately a matter of perspective.[133] For one thing, Germany did not introduce a speed limit on the Autobahn even though a speed limit of 100 kilometers per hour would have brought an immediate stop to nitrogen oxide emissions in the range of 180,000 tons per year.[134]

With the large number of unknowns, it should come as no surprise that a generous endowment for forest research was part of the response. From 1982 to 1992, government agencies supported more than 850 research projects with some 465 million deutschmarks.[135] Ecosystem research and ecological approaches were among the great winners, as forest death raised questions about the discipline's traditional focus on economic returns. Forest research also became more open for other disciplines. However, many researchers were wrestling with the exceeding complexity of the issue and the clichéd nature of "forest death" and turned to other issues when funding ran out. Nonetheless, it seems that forest research was never quite the same after forest death.[136]

Space does not allow a comprehensive discussion of all political measures, but it should be clear that the government enacted an ambitious set of responses. It probably helped that a new federal government came into power in the fall of 1982, as Kohl's coalition made the response to forest death one of its early priorities. This was particularly significant since Kohl came into office with hopes of a fundamental political realignment, a *geistig-moralische Wende*, but fears of a conservative backlash on environmental matters dissipated quickly. The new minister of the interior, Friedrich Zimmermann, was perhaps the most reactionary figure in Kohl's cabinet, but he felt no inclination to vent his political instincts on environmentalists (unless demonstrations turned violent). With that, one of the results of forest death was that it made German conservatism much greener than it had previously been, paving the way for environmental problems as mainstream issues—a status that environmental issues have retained to the present day. In fact, the swift response to forest death was probably the breakthrough for the popular notion that Germany was Europe's environmental vanguard. As Chancellor Kohl declared in a speech of 1984, "Nowhere in Europe has environmental awareness flourished so greatly as among us."[137]

All this happened in a national context, which is remarkable against the background of the aforementioned transnational initiatives in the late 1970s. The forest death debate was an eminently West German debate, and so was the response, an outcome helped by the weak results of international negotiations. When the European Community published its directive on ambient air quality standards for sulfur dioxide and dust in late 1979, the consequences were confined to some adjustments in air pollution monitoring programs.[138] As for the Geneva Convention, ratification was so slow that it did not come into force until March 16, 1983, and the first meeting of the convention's Executive Body took place in June 1983 when the Large Furnace Ordinance was about to become law.[139] The political response to forest death shows the extent to which the field of environmental policy was still firmly in the grasp of national politics.

West Germans showed an appetite for aggressive environmental policies throughout the 1980s, but this did not make policymakers fond of innovation. Interestingly, forest death did not inspire a comprehensive transformation of West Germany's environmental polity: achievements were based on preexisting tools of policy. The government acted decisively, but it never left the framework of established laws and institutions, thus making for a curious mixture of innovation and traditionalism. The government specifically refrained from the use of market-based instruments, and that was certainly not due to a lack of proposals. In the United States, environmental NGOs formed the "Coalition to Tax Pollution" and proposed a sulfur tax as early as 1971.[140] The GDR experimented with some monetary incentives in the 1960s as part of its reformist economic policy (Neues Ökonomisches System der Planung und Leitung). Among other things, GDR experts tried to establish a price for emissions.[141] The state of Hesse even produced a draft law for a tax on sulfur dioxide emissions that the Federal Council (Bundesrat), the second chamber of the German Parliament, discussed in 1983.[142] But in the end, neither environmentalists nor the government warmed up to new approaches, and the outcome ultimately confirmed what Menke-Glückert had declared when forest death was just emerging on the horizon in 1981: "The age of landmark environmental legislation is over."[143] For an environmental polity that had evolved gradually since the 1950s, forest death was the ultimate proof that it could weather a storm.

THE PERILS OF MONITORING

The German government introduced a new system of forest monitoring in the wake of the forest death debate. At first glance, it appeared to be a natural decision. As many people expected a rapid deterioration of woodlands in the near future, it seemed a good idea to monitor their health more closely. But as time went on, the endeavor became the institutional place where the paradoxical legacy of the forest death debate came into view. The results of monitoring showed a widening gap between the clichés of the forest death debate and real state of the woodlands. While the former were oozing certainty about the impending demise of Germany's forests, monitoring provided a complicated and ambiguous outlook. Forests did not perish, but they were not healthy either. The conclusions of monitoring were eminently inconclusive.

The first report on forest damage (*Waldschadensbericht*) was based on a comprehensive investigation in the summer of 1983. The German system of federalism put the forest administration into the hands of the states (*Bundesländer*), and as these states used different methods, the result was deemed insufficient. The states agreed on a common methodology before they started the next monitoring drive the following year, and the publication of the results became one of the political rituals of German environmentalism.[144] However, monitoring suffered from one fundamental problem from the outset. The underlying diagnosis was more cliché than scientific result: there was—and is—no proper academic concept of "forest death." However, there was not much room for critical remarks about popular clichés. At the height of the forest death debate, Heinrich Stratmann, the long-time head of the State Institute for Pollution Control (Landesanstalt für Immissionsschutz) of North Rhine-Westphalia, complained about a "hostility toward those who call for rational approaches."[145] Others voiced their concerns retrospectively. In 1992, Heinz Ellenberg, emeritus professor of biology at the University of Göttingen, confessed that he had harbored "doubts" about monitoring from the beginning but "had refrained from speaking out in public" because he welcomed "growing environmental awareness" in the general public and did not want to jeopardize the political response to "the growing menace of air pollution."[146]

Furthermore, monitoring could not draw on the long tradition of smoke investigations. In fact, smoke researchers were remarkably silent throughout the forest death debate.[147] They had traditionally focused on damage in the immediate vicinity of sulfur-spewing factories, and one of their prime skills was to identify the spatial range of smoke damage. But as sulfur dioxide traveled long distances and across national borders, monitoring local damage patterns was a skill of the past: pollution seemed to know no boundaries. As the Expert Council for Environmental Questions noted in a report of March 1983, "We were used to seeing the environmental toll of industrial affluence in the vicinity of large cities, centers of production, and traffic links. However, we thought that other areas were more or less unharmed. But now it seems that, at least in Europe, there is no longer anyplace that is safe from damage."[148] The underlying principle of traditional smoke research—where there's smoke, there's damage—no longer applied, and this robbed existing expertise of one of its pillars.

Against this background, scientists were usually eager to stress the complexity of the underlying problem. "All known symptoms and results indicate that forest death is an extremely complicated and multidimensional problem," a publication of 1983 declared.[149] Others chimed in, speaking of "a complicated disease" that one "could not grasp through monocausal explanations."[150] In 1986, a government research council (Forschungsbeirat Waldschäden/Luftverunreinigungen der Bundesregierung und der Länder) noted that there was "incomplete knowledge about processes in biological systems."[151] Some researchers resorted to neologisms to describe the huge variety of findings, symptoms, and approaches; one of these shibboleths was "Stresskomplex," which suggested that a broad range of factors put forests under stress.[152] But talking about complexity only got researchers so far. In 1985, Freiburg University awarded a doctorate for a study of "complexity as an obstacle to problem-oriented responses to forest death."[153]

In short, officials were not in an enviable position when they set out to build a monitoring system. While scientists embraced complexity as their mantra, administrators sought to keep the procedure as simple as possible. As monitoring was intended as a countrywide endeavor, the project involved a huge number of people, and a complicated method would markedly increase the chance of bad practices and thus undermine uniformity. Furthermore, the lion's share of monitors would come from the German forest service, where expertise on pollution problems was scarce, as the training of foresters traditionally centered on economic and administrative issues. The government eventually chose a method that focused on the extent of leaf cover—the rationale was that trees suffering from pollution would let more light through the canopy—and produced a booklet that featured pictures of trees with different levels of damage. For many foresters, it was the first time that they had a book on pollution problems in their hands.[154] Clasping that booklet, foresters all over Germany swarmed into the forests, searched for a designated tree (a uniform grid, four kilometers wide in some states and eight kilometers in others, provided them with guidance), looked up, identified what they saw as the correct one among five classes of damage, and mailed their results.

The results were disturbing in more than one way. Reports showed that damage was widespread; but they also showed that damage was different from the clichés of the forest death debate. For example, reports showed a growing extent of leaf loss on broadleaf trees, particularly oaks and beeches—whereas media reports had typically featured dying conifers. After a few years, the results also showed a significant degree of annual and regional fluctuations that defied attempts to plot trajectories for different species.[155] The situation became even more complicated when the collapse of the GDR allowed the first monitoring of East German forests in 1990.[156] With the recorded damage about twice as grave as that in West Germany, the results seemed to confirm popular assumptions about the state of the East German environment, but that argument also runs backward. Given that environmental devastation was one of the key indictments of the late GDR, monitors were certainly not in a mood to discount damage.[157]

The West German forest death discourse centered on sulfur dioxide, but the East German forests also suffered from another pollutant with a different damage profile. Large, industrial-style production units made GDR agriculture a notorious emitter of ammonia, which causes unhealthy tree growth. Nitrogen is a well-known nutrient for plants that boosts growth as no other element does, but plants also need other nutrients such as potash, calcium, and magnesium, and these nutrients tend to become deficient in the wake of massive ammonia inputs. As a result, researchers found that due to the imbalance of inputs, "forests were dying in the vicinity of nitrogen-emitting fertilizer plants and factory farms."[158] But the concept of unhealthy growth was a tough sell in a population that had grown accustomed to seeing air pollution as a killer of trees.

Monitoring showed that East German forests were recovering when air pollution decreased after reunification. While that result reinforced the reports' credibility, the development in Western Germany was creating greater headaches. Sulfur dioxide emissions decreased dramatically during the 1980s, but monitoring showed no corresponding decline in damage: the number of diseased trees was and remained high.[159] Was something wrong with the methodology of monitoring? Were the results defined by factors other than pollution? For example, Central European forests typically stand on marginal land that agriculturalists of former centuries did not turn into farmland for good reasons—was that showing in subprime growth?[160] Or were these just natural fluctuations?[161]

The inconclusive results inspired a lively debate among experts, and yet room for improvement was tight. The forest damage report was also a political event that was marked in red on the calendars of environmentalists and politicians—a perfect chance to declare one's commitment to a healthier forest and to green causes in general. That made major changes in the methodology unwise. After all, green commitments hinged on the problem's still being in the air, and a new methodology with lower damage figures would have raised suspicions of manipulation. While experts were more or less unhappy with monitoring, politicians sought to live with the ritual, making the annual press conference on the latest forest damage report a bit like the green German version of *Groundhog Day*.[162]

The perils of monitoring thus provided a fitting illustration of the cognitive dissonance that characterized forest death. As a real-world ecosystem, German forests called for a nuanced assessment that looked into local problems and potentials. Proper monitoring would have shown that the forests were neither completely healthy nor uniformly moribund; forests and pollution patterns were too complicated for either scenario. But recognizing the true complexity of Germany's forests would have raised awkward questions about popular clichés, and that made it more enticing to issue vague statements about the need to keep on the watch and not relent in ongoing efforts. The debate over forest death had led to a uniformity of rhetoric and thought that was quite at odds with the diversity of Germany's woodlands, and this made the debate a revealing mirror of the public mindset. When it came to the state of the forests, monitoring was notoriously unreliable, but the effort did show that the German population lacked a proper understanding of real forests.

CONCLUSION: THE GERMANS AND THEIR FORESTS

The forest death debate faded into the background in the mid-1980s when journalists and environmentalists moved on to other environmental issues. The 1986 nuclear disaster in Chernobyl hastened the topic's demise, and so did the Sandoz fire during the same year, which drew attention to the pollution of the Rhine and other rivers.[163] The sense of panic decreased notably over time, helped by a realization that contrary to previous fears, most German forests were still standing, but this did not inspire renewed interest. The problem still struck a nerve in the German public, but discussions no longer showed the vibrancy and the passions that typically characterize a burning topic. Experts continued to discuss the state of the forests, and bureaucrats published their reports, but German society mostly looked elsewhere. Forest death led, and continues to lead, a strange, zombie-like afterlife.

Forest death remained living memory, refreshed on an annual base through the forest damage press conferences. An artist even composed a forest death musical in the late 1990s that featured melancholic trees singing about the pain that they were incurring.[164] It had the marks of a collective effort to avoid confrontation with past errors: it was best not to talk too much about this. After all, there was no culprit that one could blame in the way that the Club of Rome was responsible for the *Limits to Growth* or Greenpeace for Brent Spar: in the early 1980s, everyone believed in forest death, save those who harbored doubts out of vested interests. It was a case of an entire country collectively moving toward an anticipation of disaster, and thus all were accomplices to the error. In any case, no real harm was done. Sooner or later, every Western country made flue gas desulfurization and catalytic converters mandatory, and the Federal Republic was just a bit faster than others. There was also no new banner message that could fuel an antienvironmental backlash. Forests continued to have a large array of problems, but they differed from the stereotypes of the 1980s. Forests were complicated and so were their problems, and complexity rarely makes headlines.

The inconclusive outcome may be intellectually unsatisfactory, but it was politically convenient. Environmentalists preferred to move on to new issues such as global warning and saw no political gains in dissecting clichés of the past. Politicians saw no potential either: environmental issues have been part of Germany's political mainstream since the 1980s, depriving critics of a context in which forest death revisionism could thrive. All the while, foresters were inclined to avoid publicity. While the forest death scare led to a flood of money for forestry and forest research, the rank and file always followed the debate with mixed feelings. While Schütt and Ulrich thrived in the spotlight, most forest experts felt more comfortable in the secluded niche that forestry traditionally occupied in German society. German forestry had good allies in state administrations, business circles, and the global research community, and they could do without politicians and journalists who probably could not tell an oak from a beech.

All this calls for caution with a view to stereotypes about Germans and their woodlands. As much as Germany displayed a peculiar attachment to its forests in the 1980s, it was a kind of postmodern awareness, with attachment to forests moderated through decades of consumerism, automobilism, and urbanization. The forest death debate did not inspire a wave of renewed interest in the real woodlands, let alone a renunciation of consumerist habits: those who had to change in the wake of the scare were power plant operators, automobile manufacturers, and industry per se, but certainly not the run-of-the-mill German citizen. In fact, many Germans had no problem differentiating between the forests under threat and the forests that they knew from personal experience. A poll conducted in 1986 showed a notable degree of sylvan bipolar thinking. A solid majority was convinced that all German forests would have perished by the year 2000—but at the same time, two-thirds of respondents admitted that they had no firsthand observations about forest death and only knew about it through the media.[165]

The forest death debate is equally remarkable for its view of the West German economy. It depicted the country as an industrial powerhouse in spite of the fact that it was well on its way toward a service economy. In fact, forest death ended up reinforcing Germany's self-image as a land of industry: aggressive pollution control was seen as a boon to the development of science and technology, which would provide the country with new markets for exports. From this perspective, industrial policy merged with environmental policy, heretofore a field without macroeconomic significance, thus producing one of the most enduring rationales for environmentalism in Germany: environmental protection was ultimately good for the economy. The new rationale found its way into programs and speeches with amazing speed: "In the long run, environmentally friendly production methods and products will open new market opportunities and create jobs," Chancellor Kohl declared matter-of-factly in 1984.[166] The boom of the service sector rarely figured in environmental debates, not to mention the ongoing realignment of the global economy and the move of industrial production abroad. From this perspective, forest death was part of self-denial about the future of German society.

None of this puts the authenticity of Germany's emotional attachment toward woodlands in doubt. Quite the contrary, the outcry shows that Germans did cherish their forests, though probably more in terms of a sphere of life than as an incarnation of Germanic myths. In any case, Germans cared about real forests, and yet the debate shows the difficult path from a general concern to a political stance: forests are complicated, and so is forest policy, and few Germans felt inclined to take a closer look even at the height of the forest death debate. In other words, a clichéd forest was probably one of the few ways to talk about sylvan issues in an urbanized society, and the sphere of forestry maintained its traditional detachment from other realms of society. It is difficult to rally political support for these issues, and when forest policy moved toward neoliberal, marked-oriented approaches in the 1990s, this happened mostly below the public radar.

It is not that environmentalists have failed to recognize the trend and stem the tide. A dozen years ago, Bavarian environmentalists made forest reform a priority and wrote a policy blueprint titled "For the Love of the Woodlands" (*Aus Liebe zum Wald*). In November 2004, 854,178 Bavarian citizens, that is 9.3 percent of the Bavarian electorate, signed up in support of a referendum. But because state law required a minimum of 10 percent of voters, the issue was never put forward for a popular vote.[167] In comparison, 13.9 percent of the electorate signed up for a rigorous ban on smoking in 2009, the referendum took place, and 61 percent of the votes were in favor of the proposal.[168] The outcome reflected the new millennium's individualism, but it also showed that while Germans continue to flock to their forests, the political windfall remains surprisingly mild.

In sum, the forest death scare changed German environmentalism and helped to bring environmental issues into a new political orbit. However, most of the political gains fell into a few years in the early 1980s: if it had not been for Chernobyl, or Sandoz, or climate change, or a broad range of other problems, a demise of green issues would have been likely. Environmentalists did not lack other topics, and nor did the Green Party, which entered parliament through forest death but soon found a whole host of other issues, including many nongreen ones, on which to leave its mark. And yet the forest death scare continues to linger. With the height of the storm more than thirty years in the past, and most protagonists dead or in retirement, a backlash has become unlikely, but so has another broad debate about the fate of the forests. It is hard to imagine another scare about the state of Germany's woodlands as long as forest death is still alive in collective memory. Forest death no longer dictates the popular discourse on Germany's woodlands, but it continues to impose limits on our sylvan imaginary—like a zombie that can bite in perpetuity and irrevocably condemns to death any discourse that it strikes. When it comes to forests, Germans have entered a postapocalyptic age.

5

THE ENDANGERED AMAZON RAIN FOREST IN THE AGE OF ECOLOGICAL CRISIS

KEVIN NIEBAUER

SINCE THE 1970S, REFERENCES to the imminent disappearance of the tropical rain forests have occupied a central place in the alarmist rhetoric of Western environmentalism.[1] Academics, journalists, activists, and sometimes policymakers, on both sides of the Atlantic have reinforced this fear with gloomy predictions about the future. Anxieties over the complex, biologically diverse, and vulnerable rain forests reached an initial climax at the end of the 1980s, after knowledge about global warming had significantly increased. At a time of accelerated globalization, the Amazon rain forest was conceived not just as a type of vegetation that was relevant to research but also as a biome of global significance. The scientific and environmentalist concerns correlated with a rapid rise in levels of deforestation in most tropical regions. In this specific context, ecological categories such as complexity, ecosystem, and biodiversity not only justified the research interests of particular academic disciplines. Scholars also linked their general doubts about the paths of human development to the future state of tropical forests by applying those concepts. Environmental alarmism relating to the Amazon rain forest was on the rise for several years until it weakened during the 1990s.

The Amazon rain forest represented both poles of critical diagnoses—its destruction could be seen as a symptom of ecological crises and its protection as an alternative to such crises. In the 1970s engagement with the issue had been primarily academic, but in the course of the 1980s it became increasingly political and popular. Local Amazonian movements, TV documentaries, conferences, articles of activist journalists, and campaigns by environmental organizations brought about this change. There were no major divergences of opinion in academic circles, as had been the case with other episodes of environmental alarmism. Most experts agreed on the causes and urgency of the problem. Scientific prognoses and data did differ when it came to detail, but these were not grounds for significant splits within the international academic community. All forecasts basically proceeded from the assumption that the Amazon rain forest would be history in a not-too-distant future.

Nevertheless, the Amazon rain forest was a highly contentious issue. Until the end of the 1980s, the discursive divide ran mainly between conservative circles in the Brazilian military government, on the one hand, and international academics and activists, on the other. The former attempted to strengthen its opposition to the latter by roping in researchers specifically for this purpose and by accusing critics of interfering in national matters. To this day, political elites, business associations and lobbyists are able to stoke fears of an internationalization of the Amazon in order to bolster their conception of progress against any critical objections.[2] The persistent marginalization and threats felt by local minorities and environmental activists are a direct consequence of this nationalist and developmentalist discourse. This is why the conflict over the Amazon has carried on bitterly until today and continues to claim many victims.[3]

In view of this enduring conflict, it is all the more surprising that the rain forest apocalypse has lost its political and public thrust. The reasons for this are not easy to determine. On the one hand, predictions of environmental catastrophes are no longer new to us, and many horror scenarios did not occur in the way they were predicted thirty years ago. On the other hand, we can see that problems that we were warned of are already beginning to materialize. Humanity is seeing with increasing clarity and frequency how extreme climatic fluctuations and the use and distribution of resources can worsen, and cause conflicts. Environmental factors are therefore considered important driving forces of migration, and recent studies have examined climate change from the perspective of national security.[4]

PRESENT, PAST, AND FUTURE

Present

At the beginning of November 2015, I came across a short, illustrated article in the German weekly magazine *Die Zeit*. According to this report, the tropical forests and the ground beneath them had been burning for weeks on the Indonesian island of Borneo. The clouds of smoke had spread as far as Singapore and Thailand. The fires had led to evacuations, public institutions had to be closed, and the archipelago found itself at the "forefront of the main agents of climate change."[5] Because of the burning peat on the forest floor, the catastrophe could not be tackled with conventional methods. In short, the situation described was undoubtedly no less serious than similar events in the 1980s. Back then, the burning forests in the Amazon region had made global headlines for several years, appearing on the front pages of well-known, high-circulation magazines, and creating concern in Western societies from research institutes to private households. At that time, Amazon deforestation had been characterized as a global inferno, an ecological Holocaust, or an environmental Hiroshima. By contrast, the mainstream media reported the catastrophic Indonesian episode of November 2015 only as a side note. Few publications approached these devastating fires in detail, and many friends with whom I spoke about the issue had not even heard of it. These observations underline the finding that burning rain forests do not affect the Western public as they did thirty years ago. This was true even in November 2015, shortly before the seminal climate conference in Paris, when it could be assumed that sensitivity toward environmental issues was at least temporarily significant. The evergreen forests of the tropics no longer inspire protesters to take to the streets in London, New York, or Berlin, and T-shirts with the once well-known slogan "Save the Rainforest" are mostly to be found in the closets of former activists. Worry about the issue can at most be activated in the short term. Interestingly, this occurs by referring to the same rhetoric, even if criticism is leveled against the media and its undifferentiated engagement with the topic. Thus, George Monbiot wrote a comment in *The Guardian* in which he reacted to the lack of media interest in the Indonesian forest fires, which he classified as "almost certainly the greatest environmental disaster of the 21st century—so far."[6] In this way, he joined the tropes of apocalyptic environmental alarmism in the 1980s. In light of these ambiguities one might raise the following question: Why do ecological threats and catastrophes such as the burning rain forests no longer provoke the same political and public influence that they did a few decades ago? Possible explanations have already been investigated, according to which, the decline in alarmism may be due to the institutionalization of environmental matters in past decades. As a result of its political successes, ecology has become something of an "idle" movement that relies too much on "established networks, arguments, and key concepts."[7]

As we will see, apocalyptic alarmism was one of the key concepts that shaped the problematization of tropical deforestation. The end of days, as a central reference in Judeo-Christian societies, is part of our cultural memory and rhetorical repertoire, and related fears can be evoked anytime there is a need to create "political pressure to prevent a crisis that seems possible or probable."[8] This cultural context partially explains why, at the height of environmental apocalypse, the future of all humanity seemed to hang in the balance in the Amazon rain forest.

Past

In the history of European colonialism the Amazon Basin played only a minor role, so that curiosity about this largely unknown space was always high. In Western societies the region was conceived as a truly different world based on Eurocentric models of thought.[9] But this frame of reference did not entirely obstruct more nuanced perspectives.[10] Thus, at the beginning of the twentieth century, the Brazilian author Euclides Da Cunha pointed out the "convenience of the 'empty Amazon' for validating observers' preconceived ideas and how important these became in establishing 'truths' about the region, how 'blankness' served to obliterate or ignore indigenous or popular history, and how external agendas diminished the ability of the observers to see what was around them."[11] The Amazon forest as a sphere of nature was characterized by an absence of civilization and culture, and the indigenous inhabitants of the Amazon were seen as belonging primarily to the natural realm.

Several reasons may explain why the Amazon became the world's most famous tropical rain forest in the age of ecological crisis. The North American dominance in the second half of the twentieth century was crucial in my opinion. The economic and political interests of the emerging superpower in the region had already begun to increase significantly by the end of the nineteenth century.[12] Over the course of the twentieth century, wealthy business-men such as Henry Ford and Daniel K. Ludwig had drawn plenty of attention to the region with their ludicrous and ultimately unsuccessful investment projects Fordlandia and Jari.[13] In both cases, Brazil had sold massive terri-tories without compunction, and what took root most deeply in the soil of these private estates was neither the rubber planted by Ford nor the *Gme-lina arborea* cultivated by Ludwig. Instead, the fastest growing plant over the following decades was Brazilian anxiety over the appropriation of the Ama-zon by foreign states, multinational corporations, and international orga-nizations. It was not just these two legendary entrepreneurial failures that made political relations with the United States particularly ambivalent. In the course of the Washington Accords of 1942, there was a growth in U.S. techni-cal missions that concentrated on research in areas like resources, infrastruc-ture, education, and health.[14] Economic cooperation between the two coun-tries remained close during the years of Brazilian military regime (1964–85), despite the occasionally tough anti-American rhetoric among political and intellectual circles.[15] From time to time, efforts were made to reduce Brazilian dependence on the northern big brother by opening its economy to other Western countries.[16]

It was also in the United States that environmentally engaged scientists, organizations, and institutions began to exert influence on public opinion and politics from the 1960s onward.[17] Overall, new environmentalism gath-ered pace most quickly in the United States, which instigated similar devel-opments in other countries.[18] North American tropical forest biologists tended to conduct research in Latin America.[19] The leading role of American environmentalism and science, together with the geopolitical constellation described above had a major influence on the way the endangered Amazon rain forest was conceived in the Western world.

Future

As mentioned earlier, apocalypticism played a central role in the conflict over the Amazon. Its political force rested on the diametrical tension that arose from the interplay of positive and negative visions of future human develop-ment. In 1940, the Brazilian president Getúlio Vargas invoked the idea of an auspicious future for the region in his seminal speech on the development of the Amazon. The rain forest should be "conquered" and its "blind force" and "extraordinary fertility" should be converted into "disciplined energy." For Vargas this was the only way the Amazon could be transformed from a "simple chapter in the history of the earth" to a "chapter in the history of civilization."[20] With this speech, Vargas created the blueprint for the developmentalist narrative of the military regime.[21] The Amazon region was seen more than ever as the ground on which Brazil's future could be built. The yet-to-be-constructed roads of the Amazon would pave the way to the *land of the future*, as Stefan Zweig had called Brazil in 1941.[22] This embeddedness in the future was largely linked to the huge scope of the Amazonian territory and its resources.[23] In the light of its continental dimensions some even compared the development of the Amazon with the manned mission to the moon.[24] Later, when the Brazilian economic and political situation was difficult, the Amazon seemed to promise a better future, as one of the world's last frontiers.[25] But Vargas had not only called for the economic development of the Amazon in 1940. He also proposed the establishment of an international research institute that would focus on the region. At the first general United Nations Educational, Scientific and Cultural Organization Conference after the Second World War, the Brazilian delegation presented these plans once again.[26] However, nationalists interpreted these efforts as a threat to Brazilian sovereignty over the region and ultimately prevented the idea from being implemented.[27] This attempt to inhibit international academic exchange on the Amazon was not particularly successful in the end, as we shall see in the following section.

ECOLOGICAL SCIENTIFICATION

After spending two years in the tropics of South America, Africa, and Southeast Asia, the British botanist Paul W. Richards asserted in the introduction to his pioneering work from 1952, *The Tropical Rain Forest: An Ecological Study*: "Owing to the activities of modern man, the area actually occupied by primary forest is rapidly diminishing."[28] Richards was one of the first of a whole generation of concerned scientists—first of all trained biologists—who over the course of the following decades would warn repeatedly of the disappearance of the rain forests.[29] The fact that into the 1960s European scientists focused their studies primarily on Africa and Asia was a result of European colonialism and imperialism, which had persisted much longer in these two regions than it had in Latin America. Not surprisingly, the African savanna was regarded as the most fascinating example of an overwhelmingly abundant and exotic nature, while the Amazon attracted only marginal attention. The scientific and aesthetic appeal of African landscapes began to decline from the 1970s onward, as can also be seen in the example of individual careers. During the 1960s, the British biologist Norman Myers worked predominantly in East Africa, until by the end of the 1970s he had become one of the leading experts on the subject of tropical forest ecology. In 1972 he drew a comparison between the African savanna and the Amazon rain forest: "Not only are these areas [savannas] fortunate in having extremely diverse wildlife—plants as well as animals—but one can easily enjoy the spectacle. I have found nothing in the rain forests of the Amazon or the cloud forests of the Andes to match the African savannah in visitor appeal, even though a tropical rain forest probably contains a greater range of intrinsic biological interest than any other biome."[30] Two significant points are made in this quotation. First, we can see that the fascination with the savanna region was related to its touristic value whereas the rain forest was appreciated for scientific reasons. Myers, in contrast to the aesthetic appeal of the savanna, saw the rain forest's ecological complexity as the ideal laboratory for objective scientific research. After having dedicated himself to wildlife conservation in Africa, Myers gave a lecture on tropical deforestation at the U.S. State Department in 1978. On this occasion he characterized the phenomenon as a "global problem."[31] This alarming speech was followed by numerous further studies, which he carried out in many cases as scientific adviser on behalf of U.S. institutions. Myers was also one of the first academics to draw a connection between the disappearance of the rain forests in Central America and beef consumption in the United States.[32] The term "hamburger connection" was coined to express the causal connection between cattle farming, rain forests, and the fast food produced by Burger King or McDonald's. Rain forest activists and scientists in the United States took up the phrase to generate public awareness of the subject. Myers also compared the rates and causes of deforestation in several tropical regions.[33] His monograph *The Primary Source*, is a highly respected standard work in the field of rain forest ecology.[34] It was significant that Myers's studies were perceived beyond academic expert circles. In the 1980s, if a North American or West European environmental organization took up the issue of rain forest protection, Myers was usually given as a reference. His studies were associated with increased support from international and U.S.

institutions, whose interest in tropical forests and the climate rose significantly during the 1970s.[35] But the Amazon gained attention not only in the
field of ecology and biology.[36] There had already been two major moments of
international outrage in earlier years with respect to the Amazon.[37] Notions
of the region ultimately depended on small scientific networks and their construction and circulation of knowledge. This is perhaps one of the reasons
that ideas and rhetoric have been characterized by continuity over a long
period. Another aspect that should be taken into account is the convergence
of international scientific-environmentalist networks and the economic
development of the Amazon in Brazil. Both processes began to materialize
between 1970 and 1972. They were initially separate from one another, since
the internationalization of environmental concern was not necessarily connected with events in the Amazon. Likewise, Brazilian developers mainly
ignored the critical objections of international experts in the early 1970s.[38]
This simultaneity of diametrical conceptions of *development* and *modernity*
augmented attention toward the Amazon. As a result, scientific and political
interest in the Amazon grew to such an extent that when "people dreamed of
'virgin forest,' they now thought first of South America rather than Africa,"
as Joachim Radkau has asserted concerning the early 1970s.[39] The importance
of that specific contextual constellation becomes evident when we consider
that the programs implemented by Brazil in the 1970s had been discussed and
planned for several years without provoking significant criticism. It was not
until that decade that the contrasts between the nationalistic discourse about
economic development and territorial security in Brazil on the one hand, and
the international depictions of global ecological interdependency, vulnerability, and crisis on the other, became evident.[40]

These contradictions were triggered by the construction of the Trans-Amazonian Highway. A heated debate arose in Brazil and abroad after the project
had been officially announced. There were disagreements not only about
whether it made sense to implement a road-construction project of such scale
but also about its social and ecological consequences.[41] Many experts considered the Brazilian government's plans to settle poor smallholder families
along the road as difficult, if not impossible to achieve. Criticism was
expressed not only in the scientific community but also in mainstream
media.[42] This was also the result of an aggressive approach of the Medici government, which touted the highway's benefits for Brazilians and international
investors. Promising announcements and print ads underlined the advan-

tages of the project, while Brazilian researchers contributed useful findings in favor of these plans. Ironically, through those campaigns the government created sufficient public awareness, which called critics to action.[43] Controversial scholars such as Henrique Pimenta Veloso, a functionary of the state's radar survey project RADAM, made absurd statements that caused a great deal of confusion.[44] He argued in favor of large-scale deforestation and said this would increase the production of oxygen while being harmless to the climate.[45] In the discussions over who had the authority and knowledge to interpret the issue correctly, it was not only intellectuals close to the Brazilian government who made false suppositions and exaggerated claims. Academics committed to conservation and ecology made mistakes as well.[46] It should also be stressed here that academic proponents and opponents of the government's plans could be found both in Brazil and abroad. Thus, the plans to use large areas of land in the Amazon for cattle farming found approval among European academics.[47] The advocates of this model of development were nevertheless a small minority. However, this did not prevent the military regime from supporting numerous such projects, particularly in the years 1969 and 1970. During the early implementation phase U.S. scholars voiced the sharpest criticism, which in turn played into the Brazilian government's hands by allowing it to instrumentalize old fears of U.S. interventionism for its own political purposes. Most scientists condemned the lack of regard for their expertise in the implementation of the development projects and a growing number of them began to apply sharper language, as they no longer behaved cautiously toward the Brazilian government. Concerned scientists also tried to achieve greater influence with their criticism by pointing out the relevance of their particular discipline. The U.S. anthropologist Betty J. Meggers expressed herself this way in the foreword of her classic study *Amazonia: Man and Culture in a Counterfeit Paradise*: "Anthropological data can no longer be viewed as curious facts of no practical use; anthropologists have something essential to contribute and an obligation to do so. If we persist in ignoring our mission, not only Amazonia but the entire planet may become an unsuitable habitat for man."[48] Beyond its task of producing scientific evidence, anthropology was also supposed to be used to avert ecological catastrophes. Meggers's study did indeed contain plenty of explosive empirical fuel, which was ignited at the right moment. Thus, her central hypothesis was that due to the fragility of its soils, most areas of the Amazon were only suitable for limited populations and specific settlement structures.[49] Most significantly, she

also referred to the government's colonization projects, since her archaeolog-
ical and anthropological findings implied that new settlers would be forced to
move farther into the Amazon after a couple of years. Based on her conclusions she made concrete suggestions to the Brazilian government, and proposed manatee and fish cultures as an alternative to cattle farming.[50] Betty
Meggers can be seen as an outstanding representative of a whole group of
academic experts who did not deal directly with the development projects in
their studies but who nevertheless increasingly addressed individual aspects
of these proposals and criticized them when necessary. By contrast, the U.S.
geographer William M. Denevan was one of the first non-Brazilian scientists
to examine and discuss the social and economic effects of the development
projects in the Amazon.[51] Denevan had responded to other studies promoting massive agricultural development in the Amazon.[52] In 1972 some biologists associated with the Mexican Arturo Gómez-Pompa published the study
"The Tropical Rain Forest: A Non-Renewable Resource" in the journal *Science*. The authors of this study stressed the ecological complexity of the
biome while pointing out that there were still many scientific shortcomings.
They also problematized one further aspect that up to then had received only
little attention. According to that, Gómez-Pompa and his colleagues questioned whether the Amazon forest would be able to regenerate itself under
the current practices of land use. Using that argument the authors expressed
their worries about the future existence of their object of research: "Even
though the scientific evidence to prove this assertion is incomplete, we think
that it is important enough to state and that if we wait for a generation to
provide abundant evidence, there probably will not be rain forests left to
prove it."[53] Here, concerns about the urgency of the matter necessarily had to
take precedence even over scientific evidence in order to prevent a massive
loss of species. In the same year, at the Stockholm Conference, developments
in the Amazon were already labeled as an *ecological Hiroshima* even though
affected areas were relatively small at that time.[54] However, from a scientific
point of view, the as yet uncharted biological diversity of the tropical forests
was already sufficient grounds for warnings against further deforestation.
Proponents and opponents of Brazil's Amazon policy seemed to be particularly united in their conviction that the government and the institutions
involved lacked qualified personnel and technical expertise.[55] There was also
a consensus that too little was known about the Amazon rain forest and that
it represented a "vacuum of science."[56] This objection ran the risk of continu-

ing the centuries-old idea of the "empty Amazon."[57] In that respect indigenous knowledge and practice was not considered conducive to modern science. Not surprisingly, Western academics were not primarily preoccupied with the local cultural scale of rain forest crisis in times of accelerated globalization. Instead, the home of the indigenous people was conceived as a field of research of global significance. Ecological categories such as *complexity*, *ecosystem*, and *biodiversity* not only justified the research interests of conservation biologists but also underpinned the personal doubts about Brazil's Amazon policy.[58] Those doubts were usually expressed in the forewords of scientific studies. Some biologists stood out when it came to the problematization of the social and economic situation on Amazonian ground. One of them was the U.S. ecologist Philip M. Fearnside. After long periods of research in the Amazon, he remained permanently at the Instituto Nacional de Pesquisas da Amazônia in Manaus, and he was not afraid to couple empirical findings with political demands.[59] Another important early academic critic was the German biologist Harald Sioli. From the 1930s to the 1950s, he had lived in Brazil, where he explained the nutrient deficiency of Amazonian soils on the basis of limnological studies.[60] In doing so, he was one of the first to contradict the widespread assumption that the Amazon region was particularly fertile and suitable for modern agriculture.[61] In an interview with the Brazilian magazine *Realidade* in October 1971, Sioli commented on the atmospheric effect of tropical rain forests. The popular view that the Amazonian forests are the "green lungs of the planet" was later drawn from this explanation, reframing the statement he actually made.[62] Sioli had simply warned that the deforestation of the rain forests would lead to a rise in the concentration of carbon dioxide in the atmosphere, and did not mention the fact that a forest "uses as much oxygen as it produces in the balance between growth and decomposition of organic substance."[63] Nevertheless, this catchy and memorable metaphor was essential for several years when it came to rain forests and climate change, not least because it served scientists and activists as a means of applying pressure and simplifying a complex issue. Another catalyst in the ecological scientification of the Amazon rain forest was the growing interest of the media in the environment from the 1960s onward.[64] In the face of Cold War tensions, oil crisis, economic recession, and growing unemployment rates a sense of crisis accompanied by a "noisy concern for the future" evolved in the United States and West Germany during the 1970s.[65] This was the context in which scientific prognoses on the future of

the Amazon rain forest became an important tool in attracting media attention and exerting political pressure.

Future Prognoses

First of all it has to be made clear, that the alarmist mode of a growing number of experts was not unjustified. Deforestation rates in the Amazon were
clearly rising since the 1970s. Between 1978 and 1987, 21,130 square kilometers of primary forests had been deforested in Brazil annually.[66] Between
1980 und 1989 absolute deforestation numbers in the Brazilian Amazon were
the highest on a global scale.[67] The rise of prognostics as a tool of concerned
scientists was also grounded in epistemic and technological developments.
Altered knowledge practices, technologies, and modes of thinking paved the
way for political initiatives as they objectified and rationalized environmental
problems.[68] Needless to say, recommendations for political action were not
part and parcel of a natural scientist's life. As a result, concerned scholars had
to find a strategy to combine their growing wish to sensitize the public with
their self-understanding as respectable scientists. An episode involving two
U.S. botanists, Robert J. A. Goodland and Howard S. Irwin, demonstrates
this well. The construction of the Trans-Amazonian Highway was the trigger
for their study *Amazon Jungle: Green Hell to Red Desert?* With a view to the
possible social and ecological consequences, the authors called for the assessment of the project to be coupled first and foremost with academic research
on the Amazon.[69] As an alternative to building a transregional road, they
advocated an increase in water transportation. The study created a big stir in
Brazil and was harshly criticized and censored by the government, leading to
a dispute in 1976 over the Brazilian edition of the book. The publisher at the
renowned Universidade de São Paulo had evidently not included the introduction and a critical chapter about the relationship between science and the
indigenous people in the Portuguese version. The Brazilian environmentalist
José Lutzenberger, who was familiar with the original edition of the book,
noticed the omissions. He turned to some Brazilian newspapers to make the
matter public. Furthermore, he asked the two American authors for permission to translate the missing passages himself so that he could distribute
them with the help of his environmentalist organization Associação Gaúcha
de Proteção ao Ambiente Natural. The South Brazilian newspaper *Jornal da
Tarde* reported on the conflict and printed an extract from the authors' intro-

duction, which had been censored and was now translated.[70] Goodland and Irwin were among the first foreign researchers to conduct studies along the new axes of development in the sense of an "applied ecology."[71] It is already evident from the title of their book that the authors were aiming at political and public influence. In doing so, Goodland and Irwin were reacting first and foremost to the approach of the Brazilian government, whose overly optimistic outlook for the development of the Amazon involved giving little consideration to negative socioecological consequences. As had Gómez-Pompa three years before, they feared that the complexity and vulnerability of the Amazon ecosystem probably meant that it would be unable to recover after widespread deforestation: "All that has emerged from the not inconsiderable scientific research and the diverse practical experience of man in Amazonia indicates that the ecosystem of the hylea is among the most vulnerable on our planet. Poorly endowed with nutrient reserves to support plant growth, its continued productivity depends on the permanence of this most complex of forest ecosystems which, once destroyed over extensive areas, will never re-establish itself."[72] Concerned scientists by no means expressed themselves as drastically as Goodland and Irwin, but they did offer alarming predictions about the future more frequently. A warning was given at the Stockholm Conference that the opening of the Amazon might not only "severely endanger the ecological balance of Brazil, but of the whole world," as it was picked up by the German magazine *Der Spiegel*.[73] From then on, practically every study that was critical of developments in the Amazon generally included two components: an approximate moment in the future when the rain forest will have disappeared and an idea of the future condition of the endangered biome. Also in 1975—the year Goodland and Irwin published *From Green Hell to Red Desert?*—Robert Allen, a coauthor of the *Ecologist*'s "A Blueprint for Survival" (1972), warned in the *New Scientist* that "all the world's tropical rain forest, apart from a few scientific mementoes, will be destroyed within 20 to 30 years."[74] Taken as a whole, the projection on the future of the Amazon was a risky approach considering that the data available on patterns and scales of deforestation were imprecise and remained disputed for many years to come.[75] Nevertheless, during the 1970s several scholars assumed that most of the rain forests would already be history by the beginning of the twenty-first century, whereas in the following decade, those estimates had to be revised as it became clear that early predictions did not prove right despite constantly rising deforestation rates. Yet most researchers were more alarmed

than previously by the dynamics of tropical deforestation, which remote sensing via satellite had made more evident. Some forecasted a "wasteland" as in
the arid Northeast of Brazil.[76] Others even feared a "second Sahara."[77] In *Science* the geologist Irving Friedman warned that the Amazon could probably
turn into "another Sahel."[78] The renowned German zoologist Ernst J. Fittkau
drew an alarming picture of "leached-out deserts."[79] In rare cases, scholars
referred to the Amazon region itself to illustrate their concerns, though it
seemed less appealing than referring to topoi like the Sahara and Sahel. For
example, Ludwig Beck warned that the rain forest could turn into steppe just
as it had already in the *Zona bragantina* close to the Amazonian city Belém.[80]
U.S. experts shaped the scientific discussion about the endangered Amazon
in the 1970s and 1980s, not least because of their large number and considerable financial support from national research programs. Brazil fit most easily into the role of a reactionary, authoritarian, and unteachable emerging
economy. In the eyes of many Western researchers, the government acted like
an "enfant terrible" that was stubbornly taking bold steps with no regard for
rational considerations.[81] In the 1980s, scholars more frequently mentioned
the social and cultural implications of the transformations taking place. The
indigenous people of Brazil increasingly took part in the discussions and
they were supported by Western activists and scientists. International back-
ing for their struggles against further deforestation grew strong, and indigenous demands became a highly sensitive and controversial area of conflict in
Brazil. Nationalist politicians and intellectuals reacted by blaming and crit-
icizing concerned scholars, whereas some of them responded by drawing a
clear line between *serious science* and *emotional environmentalism*: "In the
view of many observers—switched-on scientists, not 'wildlife eco-nuts'—the
end of the century could see much of the biome reduced to degraded remnants, if not eliminated altogether."[82] Myers's assertion aimed at the epistemic
authority and higher validity of scientific knowledge. Only scientists like him
could make reliable statements on the future of tropical forests, although the
boundaries between science, politics, and environmentalism had increasingly blurred in light of a contested issue, as we will see later.

Tropical Deforestation Seen from Above

While there was a general consensus about the urgency of the matter, on
a technical level it was difficult to carry out scientific examinations. It was

hard to capture and depict deforested areas, which were quite often located in remote areas that were difficult to access. Scholars could spend their whole lives measuring deforestation in the vast territories of the Amazon. The scale of the Amazon was a significant problem for researchers, the Brazilian government, and environmentalists alike. Against this background it is evident that from the 1970s onward, remote sensing via satellite became the most important technology in the field of tropical deforestation. This new form of data collection allowed many processes to be visualized and recorded for the first time in history. The Landsat Program was initiated by the U.S. National Aeronautics and Space Administration in 1972. Several generations of Earth-monitoring satellites have orbited the Earth since then. Countless maps attaining resolutions of 15 meters per pixel were produced during that period.[83] Though this technology was relevant to all kinds of stakeholders, above all, it strengthened the position and argumentation of concerned academics.[84] Remote sensing was supposed to significantly reduce uncertainty in measuring and visualizing the problem, which strengthened the epistemic authority of scientists.[85] The technology made it possible to reconstruct and substantiate the extent of forest clearings, at least on a rough scale. The satellite perspective also matched the self-conception of academic experts and their idea of objective science-based evidence. The rain forest measured from space could be conceived as a "manageable object."[86] Through satellite imagery, findings on the history, present, and future of the Amazon could be interrelated in a new way. The data thus generated became important for environmental organizations because they could exert political pressure with "hard facts." Mainstream media like *Der Spiegel*, *Time Magazine*, and the *New York Times* quickly adopted the prognoses, while for understandable reasons, few concerned scientists mentioned that remote sensing also exposed vulnerabilities and that uncertainties could not be avoided.[87]

Not only pictures taken from space but also images captured from lower levels of the atmosphere became important visual vehicles for conveying this new understanding of the Amazon. Elegantly curved river landscapes amid dense rain forests with clouds appearing and floating across the forest canopy depicted the unspoiled paradise from an aerial perspective in the increasingly important field of nature documentaries. The negative, contrasting images revealed burning forests covered by billows of thick smoke. When the images were taken at night, the region could be visualized as an accumulation of countless tiny points of light against a black background. Remote sensing allowed more than seven thousand fires to be counted across the whole Amazon region on September 9, 1987.[88] Numerous airports in South America had to reroute their flight paths due to smoke carried by the wind, and for months journalists and environmentalists referred to these disturbing images and to the data extrapolated from them to illustrate the extent of deforestation. Seen from above, the burning Amazon rain forest was a measurable tropical inferno.

POLITICIZATION, INSTITUTIONALIZATION, AND POPULARIZATION

If the central discussions in the 1970s had been among academic scholars taking issue with developments in the region, in the next decade the spectrum of actors widened. In particular the period between 1985 and 1992 brought the most important changes with regard to the Amazon. It became an area of conflict between politicization, institutionalization, and popularization, as exchanges between different arrays of actors intensified significantly. Journalists, academics, and environmentalists from various countries met regularly at conferences and exchanged views and information. They usually shared similar concerns and spoke the same language when it came to the endangered rain forest. The respective discourse communities can therefore not be clearly reduced to their professional or national backgrounds, as they formed a specific epistemic community. Some additional factors played an important role in giving the Amazon such outstanding importance. The intergovernmental and geographical proximity between the United States and Brazil favored the former's interest in the region. This also affected Western Europe, since European researchers and environmentalists were particularly oriented toward their colleagues across the Atlantic. The fact that other tropical forests attracted much less attention from the Western public, despite similar or even worse processes of transformation, attests to the significance of these primarily transatlantic networks. Between 1980 and 1990, some areas of Southeast Asia and West Africa had significantly higher rates of deforestation than the Amazon.[89] Compared to the original extent, primarily the West African rain forests suffered from the most dramatic decline since the 1970s, and the Western public took not much notice of it.[90] It may be that the Amazon region garnered more attention than other tropical forests because of its unrivaled scale, fluvial topography, water reserves, and biodiversity. References to the superlative nature of the Amazon were and still are a recurring argument made by those concerned with deforestation. Nevertheless, these character-istics do not sufficiently explain the outstanding importance of the Amazon during those years. Unlike the forest death debate in Germany, no generic term shaped the problematization of the issue.[91] Instead, the trope served as a complex reservoir of various concerns and attributions. As had the forest death debate in Germany on a national level, the *last great rain forest* had a binding effect on diverse sectors of society.[92] One important difference between the two tropes was that the concerns about the endangered Amazon worked across national boundaries.

Politicization

Politicization can initially be traced back to the growing importance of social movements in the Amazon. Local actors, who up to this point had been mostly excluded from the international debate, were now in more inten-sive contact with researchers, policymakers, and institutions, and they drew increased attention to the social, cultural, and economic aspects of tropical deforestation. Significant change came with the political visibility of local Amazonian activists from 1985 onward. The Brazilian anthropologist Alfredo Wagner Berno de Almeida has described this process as an act of "separation from nature."[93] Even if this argument seems convincing, it should be added that this separation did not occur as radically as Wagner de Almeida suggests. Both the indigenous activists and the rubber tappers of the Amazon had some difficulties during this period in working out how to position their ways of life and political claims in the conflict between ecological-international and economic-national oppositions. They tended to be oriented toward ecologi-cal thinking, even though many of them wanted to differentiate themselves more clearly.[94] In so doing, conflicting or contradictory self-portrayals of indigenous leaders were not uncommon, as they referred to Western eco-logical principles on the one hand, while at the same time trying to liberate themselves from the essentialized qualities that environmentalism attributed to them. The persistence of Rousseau's idealized notion of the "noble savage" could be decisive in determining which indigenous groups received politi-cal and financial support and which ones were denied it.[95] The latter mostly affected those who, because of their advanced acculturation, no longer corresponded to the ethnocentric conceptions of their culture, even though they still possessed and passed on indigenous knowledge and practices. From the perspective of many environmentalists and scientists, they were primarily a warning example of rain forest destruction and cultural decline. Cultural adaptation could reduce their right to protection in competition with other *more traditional* groups. The dialectical logic of Western rain forest alarmism reinforced these conflicting ascriptions.[96] Thus, it was generally very difficult to define a common agenda in the field of tension between state institutions, missionaries, international NGOs, and their own organizations. In the worst case this could result in splits within the community.[97] It was therefore all the more astounding that activists such as Chico Mendes managed to dismantle the essentialism of Western conservationism by appropriating the underlying ecological categories and combining them with the political claims of the rubber tappers. In this way, their local socioeconomic conflicts and the global ecological implications of the endangered rain forest could be problematized and negotiated. Although from a sociocultural point of view there was often a big gulf to bridge between locally oriented activists in the Amazon and environmentalists from North America and Europe who thought in global terms, this cooperation worked astonishingly well.[98] The integration of different views and interests seemed feasible with respect to the rubber tappers because they could not be clearly placed into the dualistic opposition between *nature/ tradition* and *culture/modernity*. But even in Chico Mendes's case, the alliance between international environmentalism and the local political struggles did not proceed without contradictions and tensions.[99] One consequence of these solidarity coalitions was the concept of the "reservas extrativistas." This system of land use allowed the rubber tappers to define a more socially just, ecologically sound, and economically productive method of using the resources of the tropical forest. The extractive reserves were researched by allied scholars, publicized by the media, and at least perceived by politicians.[100]

Flora and fauna, biodiversity, and the ecosystem may have been the main areas of interest for scientists and environmentalists up to this point, but social, cultural, and economic implications of tropical deforestation were increasingly taken into account. Accordingly, beginning in the 1980s social scientists shaped the debate on the endangered Amazon more than they had in the previous decade.[101] At the same time, it is not surprising that prognoses were still widely used as a mode of political pressure given the rapidly rising rates of deforestation.

Institutionalization

The socioeconomic and ecological consequences of Brazilian development projects were also discussed in national parliaments, expert commissions were set up, and multilateral agreements were signed. As awareness of climate change grew, the Amazon rain forest became increasingly conceived as a regulatory space of societal relevance in countries like the United States, West Germany, or the United Kingdom.[102] Several conferences were organized during these years on both issues. Environmental concerns and worries about the future were addressed at the highest political levels, as demonstrated by the extensive *Global 2000 Report* commissioned by the U.S. president Jimmy Carter in 1977.[103] Two years later, the U.S. Interagency Task Force on Tropical Forests was set up to promote the protection of tropical forests on a global scale.[104] From the mid-1980s onward, the topic was regularly on the agenda of the U.S. Congress. In 1986 the Tropical Forestry Action Plan was created under the direction of the Food and Agriculture Organization of the United Nations, the World Resources Institute, the World Bank, and the United Nations Development Programme. The objectives of this five-year plan were reforestation as well as the transfer of technology and expertise, aimed at maintaining a balance between conservation and development. The Study Commission set up by the 11th German Bundestag in 1987, "Protecting the Earth's Atmosphere: An International Challenge," dedicated a whole chapter of its interim report to the protection of tropical rain forests.[105] The commission discussed the status of research on causes and effects of global deforestation and highlighted measures that could be or had already been taken. In 1990 the commission published a comprehensive report entirely on the topic "Protecting the Tropical Forests."[106] References to the Amazon's importance in regard to global climate now formed a central argument in addressing the problem. The first NGOs dedicated to working exclusively toward rain forest protection were established in 1985–89.[107] Even conservative European politicians who were not particularly well-known for their environmental agendas expressed their concerns at international conferences in the late 1980s.[108] Richard von Weizsäcker, president of West Germany qualified rain forest politics as "a question of human survival."[109] At least in countries like the United States, Germany, and the United Kingdom, tropical deforestation was seen as a global-scale problem across the political spectrum. Despite this whole process of politicization and institutionaliza-

FIGURE 5.1. The *Ecologist* Magazine, 1989, "Amazonia: The Future in the Balance."

tion, the Amazon remained an icon of untouched wilderness and premodern closeness to nature.[110]

Popularization

As a result of the two processes described above, the public in the United States and West Germany began to perceive the endangered Amazon rain forest as a symptom of ecological crisis that was likely to affect the whole globe. This was also related to climate change, the greenhouse effect, and ozone layer depletion at the end of the 1980s. In being linked to these phenomena, the endangered Amazon rain forest was thematized and visualized in school notebooks and children's magazines, on cereal boxes, stamps, and posters. In Germany the quest for rain forest survival was even simulated in a board game. Iconic animals such as the jaguar, ocelot, toucan, or red-eyed tree frog highlighted the fascinating nature of this biome, its biological diversity, and the need to protect it.

PERSISTENCE OF THE ENDANGERED AMAZON RAIN FOREST

Most of the once powerful tropes of ecological alarmism have become permanent background noise, despite the continuation of manifold crisis phenomena. This is also the case for the endangered Amazon rain forest. After the issue had gained importance, especially during 1985–92, its societal relevance in Western countries swiftly faded, in particular when it comes to the intensity and regularity of media coverage and, similarly, to the degree of public awareness. The narratives and imagery produced by those involved in this area of conflict have hardly changed since the period under consideration in this chapter. Meanwhile, scholarly and environmentalist engagement with the subject has continued in times of increasing specialization and institutionalization. Consequently, the exchange between and intersection of different actors and discourse communities has declined since the heyday of the rain forest apocalypse. All these findings lead us to the question of the legacy of the endangered Amazon rain forest.

In 2013, the Brazilian environmental historian José A. Pádua traced the reasons for declining rates of deforestation in the Amazon since 2004.[111] He and others have attributed this development partly to the policy of the former minister of the environment Marina da Silva (2003–7).[112] However, Pádua also referred to another factor that is central to the issues described here—agricultural production in the great savanna (Cerrado) of central Brazil has likewise expanded in the past few decades.[113] Because the Cerrado region has never attracted much attention, the agricultural sector was able to establish itself there almost without raising any major concern.[114] Furthermore, the intensification of agribusiness in the Cerrados is connected with the rise of the endangered Amazon rain forest as a transatlantic trope of the ecological crisis. With respect to some factors discussed above, the two regions reveal similarities when compared to each other. Due to the Cerrado's rich biodiversity it could likewise have been in the spotlight. Just as in the Amazon, large-scale agricultural expansion had begun in the 1970s—that is, in parallel with growing environmental awareness.[115] For tourism purposes it might have been more suitable than the tropical rain forest, given its favorable climate and its geographical proximity to the urban centers in Southeast Brazil. As we can see, the academic and environmentalist focus on the Amazon ultimately had concrete consequences for the direction taken by the Brazilian agricultural sector. This is where the implications of environmental alarmism come to the fore.

Today, extractive reserves can be found across Brazil, most of them located in the Amazon region. These protected areas can certainly be seen as the legacy of Chico Mendes and his many allies. Apparently, there is a remarkable degree of persistence here. Although the network of Chico Mendes only existed temporarily, he is still a well-known figure referred to on both sides of the Atlantic in regard to rain forest protection. He is even known beyond the Western hemisphere, and a conference was held in Washington, DC, where his former circle of supporters gathered on the occasion of the twenty-fifth anniversary of his murder.[116] On October 15, 2015, a symposium was organized at the University of Brasília to commemorate the first national meeting of Brazilian rubber tappers thirty years earlier. The meeting was held at Praça Chico Mendes (Chico Mendes Square).

Nevertheless, public and environmentalist concern about the Amazon has decreased since the mid-1990s and it is difficult to clearly assess the contemporary situation. On the one hand, contrary to most prognoses, large portions of the Amazon rain forest are still standing, and over the past twenty years considerable parts of it have been placed under protection. Most other tropical rain forests on the planet are in much worse condition. On the other hand, even in the Amazon there is no evidence of a long-term easing of the situation, either in a political or an ecological sense.[117] Some fears with respect to the ecological correlations of the Amazon have become apparent at a regional level, for example, in the 2015 water crisis in Southwest Brazil.[118] Finally, in 2013 the deforestation rate grew by 28 percent compared to the previous year, after having dropped steadily for seven years.[119] Between 2000 and 2010, absolute deforestation in the Amazon rain forest still made up 44 percent of all tropical deforestation around the world.[120]

Today, when the issue is addressed beyond specialized academic circles, the images and narratives that gathered strength in the 1980s are still referenced. Behind these representations we find, in part, centuries-old Eurocentric tropes of tropical nature. Maybe their durability relates to the difficulties of contemporary societies in keeping up with the rapid and profound transformations taking place in the Amazon and beyond in times of the "Great Acceleration."[121] For all these reasons, the endangered Amazon rain forest is characterized primarily by persistence, even though alarmism has faded into the background.

6

GREENPEACE AND THE BRENT SPAR CAMPAIGN

A PLATFORM FOR SEVERAL TRUTHS

ANNA-KATHARINA WÖBSE

IN THE SPRING OF 1995, the German evening news looked like scenes from a how-to guide for effective environmental campaigning. It showed pictures of police leading away Greenpeace activists from a rusty oil rig somewhere far out in the middle of the North Sea. This was just the start of one of the most important Greenpeace campaigns that Europe had ever seen. A veritable media storm erupted over the following weeks, as campaigners deliberately fed it with stunning images that left a lasting impression on the visual canon of environmental memory culture. Millions of people saw flimsy dinghies with activists in full-body protection suits under attack from water cannons against the backdrop of the disused Brent Spar oil rig, which Shell planned to dispose of in the ocean.

The pictures became a point of reference in the environmental discourse. A massive loading rig for North Sea oil, about to be broken up and dumped into the Atlantic ocean, became a powerful symbol encompassing several dimensions. The confrontation served as an icon for the environmentalists' resistance to the disposal practices favored by industry. In fact, Brent Spar stands as one of the largest boycotts of an international corporation to date. It was also a campaign that achieved its goal, insofar as Brent Spar was not dumped into the sea as planned but disposed of on land according to strict protocol. But as the campaign lingered in collective memory, it emerged that Greenpeace had for some time argued using false figures: the organization had suggested that the rig held far more toxic material than it did. That gave the campaign a different twist—it misled the public, at least for some time. First and foremost the media came to reflect on whether they had let themselves get carried away by Greenpeace. In short, while some came to see Brent Spar as the epitome of civil society's resistance to the hubris of multinational oil companies and politics, others viewed the campaign as emblematic of ecological and media hysteria.

All this happened some twenty years ago. With the facts mostly undisputed, it is time for an assessment of the event's place in collective memory. How do the different storylines look in retrospect? Specifically, how does knowledge of the dubious factual statements change our assessment of the event's place? Do Greenpeace's scenarios belong to the category of apocalyptic storytelling? Was the event even a false alarm? Did their dubious arguments cast a long shadow on these actors? In reviewing contemporary documentation, it becomes clear that this case had an extraordinarily high level of media attention. Few actions or campaigns have had so much coverage. There was also a great amount of academic interest, particularly from media studies and political communications. This article focuses particularly on the German dimension of this story, as Germany had by far the most vigorous response to the Greenpeace campaign.

REALITIES ANCHORED IN THE GROUND

The North Sea emerged on the world map of oil production in the 1960s when deposits of crude oil and natural gas were discovered in quantities large enough for investors. In 1971, two oil companies, Exxon and Shell, started to exploit the so-called Brent Field: "The North Sea oil rush had begun."[1] The oil crisis of 1973 and the quest of Western countries for domestic reserves supported the boom. Oil companies invested heavily in mining equipment and technical infrastructure.[2] Along with Norway and the Netherlands, Britain stood to profit from the bonanza. Within only ten years, supplies from the North Sea oil fields made the country effectively self-sufficient in regard to fossil fuels.[3] As for interests, obviously politicians and the oil industry had some common ground.

Brent Spar itself was not a drilling platform, but a storage and loading platform built to ease the safe transport of raw material by tankers from the oil fields back to the Shetland Islands. Shell UK Exploration and Production (Shell Expro), a joint company of Shell UK and Esso UK, owned the rig.[4] The offshore facility Brent Spar resembled a huge buoy with a diameter of 29 meters, held in place by 6 chains to the floor of the North Sea. It was 137 meters tall, but with a full load, only 28 meters were visible above water. This part included the machine deck, workers' accommodations, and a helicopter pad where cranes and loading machinery for filling tankers could be installed. The rest of the structure remained below the surface. The full weight of Brent Spar was about 14,500 tons. In size and weight it was similar to a large channel ferry.[5]

In 1976, Brent Spar was assembled in Norway, brought to the Brent Field in an upright position, and anchored some 190 kilometers to the northeast of the Shetland Islands. Brent Spar's role in storage and loading was diminished with the completion of a pipeline to the Shetland Islands in 1978. Nonetheless, Brent Spar was left in place as a backup in case of difficulties with the pipeline. In 1991 the Brent Spar was shut down after fifteen years of use. The oil tanks were emptied and filled with seawater, the pipelines were washed out, and the remnants placed in a tank. Brent Spar sat unoccupied and non-operational, waiting for disposal. Only one thing was left to do: find the best time and way to get rid of it.

According to Shell Expro, it commissioned more than thirty examinations and studies. Experts from consulting firms and universities sought to identify the "Best Practicable Environmental Option."[6] This process was in line with British law. For the process of waste disposal, Shell needed the government's permission. Brent Spar fell within the EEZ (Exclusive Economic Zone) of Great Britain, which in turn was bound by international law. In the end, Aberdeen University Research and Industrial Services was given the task of selecting the best solution from six options, which included cleaning, repairing, and selling it, dismantling the rig, or taking it back to Norway.[7] The experts voted for disposal in the deep sea, a decision that the engineering firm Rudall Blanchard confirmed in its final assessment by labeling it the "best practical environmental option." Moreover, disposal at sea would cost only £11 million whereas the price tag for land disposal came to almost £46 million.[8] Shell opted for disposal at sea and filed the papers with the British authorities.

Legislation in this field is extremely complex.[9] As early as 1958 the United Nations assembled a Convention on the Law of the Sea to define some terms

of use for shared international waters. On the one hand, the agreements recognized the right of coastal sovereign states to mine the mineral resources in
their territorial waters. On the other hand, they also required the signatory
nations to ensure that installations were fully removed after use.[10] This statement shows that surprisingly little thought was given to disposal when oil
rigs were being built. Specifically, the large offshore rigs in depths of more
than seventy-five meters were evidently constructed without any plans for
their deconstruction. British lobbying led to a watering down of the mandate
for complete dismantling, and this more ambiguous mandate was enshrined
in the Convention on the Law of the Sea in 1982. The agreement stipulated
that dismantling should adhere to "generally accepted standards established
in this regard by the competent international organization." Oversight on
this agreement fell to the International Maritime Organization (IMO), which
found that structures had to be removed completely except in "special circumstances" that were to be assessed and approved on a case-by-case basis.[11]
Writing a few years before the Brent Spar incident, an insider offered the following outlook: "Complete removal may become the norm for the southern
North Sea, non-removal may be accepted for a few of the larger installations
in the northern North Sea, particularly where there are concerns over the
technical feasibility or hazards involved in dismantling or removal operations." In short, it firmly established the dumping of facilities into the sea
as an acceptable method of disposal within the business community before
Brent Spar.

In 1988, the United Kingdom Offshore Operators Association wrote a
report outlining the future disposal costs of facilities on the UK continental
shelf. The organization concluded that dismantling all facilities would cost up
to £4.4 billion. However, a policy of partial dismantling and optional dumping of selected facilities might save £1.5 billion. The matter pertained directly
to the government's finances, as most disposal costs for oil companies were
tax-deductible. In short, the higher the disposal costs, the lower the government revenue.[12]

The matter ultimately went to the Department of Trade and Industry,
and the licensing procedure ended in February 1995. The platform would be
sunk during the summer at a spot in the East Atlantic where water was two
thousand meters deep. Conservationists in Scotland argued along the lines of
Shell that disposal of the rig in deepwater would have negligible local effects.
However, Greenpeace took a different stance, and its German *Greenpeace*

Magazin published its first article on Brent Spar in February 1995. Its glaring headline read: "Here today, gone tomorrow—an old oil rig is being dumped in the sea."[13]

At this time, Greenpeace was preparing a campaign on the occasion of the North Sea Conference, which was scheduled to take place in the Danish city of Esbjerg in June 1995. Marine protection had always been at Greenpeace's center of activity since its establishment in 1971.[14] The German branch of Greenpeace moved into the public spotlight in October 1980 when activists blocked a ship owned by the Kronos Titan chemical corporation and used for the dumping of acid wastes in North Sea waters.[15] The North Sea had been on the environmental agenda long before Brent Spar. The large inputs of toxic and harmful substances via rivers and the discharges of chemicals, nuclear waste, and other residues caused visible and even irreversible damage to the North Sea ecosystems. In Germany, the Council of Environmental Advisors' "Special Report on the Environmental Problems in the North Sea" gained widespread attention in 1980.[16] In 1984, the first ministerial "North Sea Conference" had already met with vigorous protests from environmental activists. A transnational alliance of environmental and natural protection organizations founded the "Action Conference of the North Sea," and the group criticized the weak and nonbinding agreements that the meeting produced. The symptoms of a maritime ecosystem under severe pressure piled up: in 1988 thousands of seals all along the coasts died from a virus. An educated guess suggested that the animals were in poor condition due to their polluted habitat. The dead bodies on the beaches seemed to forecast a dark future for the maritime sphere. During the following years, the North Sea conferences received widespread recognition, and representatives of the German government sought to present themselves as staunch advocates for the protection of the high seas.[17] Yet the political negotiations ground on and environmentalists waited for tangible results. The Brent Spar campaign took place against this background. Greenpeace set out to accelerate things.

Greenpeace International's chemical activists sought to use the North Sea Conference to expose the problem of chlorine-based industrial chemistry, one of the worst maritime polluters in the area, and they searched for a powerful visual hook. A Dutch Greenpeace campaigner, Gijs Thieme, identified the key obstacle to campaigning for the protection of the North Sea: "A North Sea campaign suffers from the invisibility of the danger and the anonymity of the perpetrators." Politicians do not react unless they come under massive public pressure: "And we can only build pressure when there's a symbol, which the political message can be associated with and which it can set alight." Thieme neatly captured his conception of a campaign: "It is always symbolic work, simple confrontation, with significant depth." This rationale captures the tendency of environmental campaigns to have a certain alarmist undertone as they inevitably sketch scenarios based on idleness. Only action can put a halt to such a negative development. Thieme heard about the plan to dispose of offshore oil rigs in January 1995 from his British Greenpeace colleagues and about the particular case of Brent Spar: "I had a gut feeling that I had hit on something important. A prominent symbol, a prominent perpetrator."[18]

This "gut feeling," the emotional awareness of the potential for public engagement was presumably based on years of experience in campaigning. It was a strategy to shrink specific issues and problems into an interesting, media-compatible format with the Greenpeace logo prominently displayed. The backdrop for this type of "product placement" was the rusty oil rig, against which the brave but defenseless activists would make a scene. It was protest custom-made for the media: "The visibility of the cause of conflict, and inequalities between the opponents are imperative in an age of audiovisual media: conflicts that can only be solved through negotiations at international conference tables receive widespread attention only when they are paired with activist strategies with their dramatic stages, which, like sports contests, can be constantly followed by viewers from their own home."[19] At its inception, the campaign was destined to work within the organization's broader North Sea campaign. However, Thieme met with skepticism internally. For one thing, there was no budget and no staff capacity for such a "jaunt"; furthermore, the meaning of such actions, the inner core of Greenpeace campaigns, was a subject of internal debate at the time. For some, the potential costs involved were too high; others were concerned about focusing too much public attention on Brent Spar. And finally, some believed that "classic actions—and Brent Spar was one of these—were outdated from a policy and organizational standpoint."[20] It fell to the action division, which was subordinate to, or at least on a par with the campaign departments, to lobby for the Brent Spar campaign internally. On April 11, Greenpeace decided in favor of occupying the rig from May 1 until the end of the North Sea Conference in June 1995.

The environmental organization Greenpeace has been "an expert in campaigns" ever since it was established in Canada in 1971. Unlike other environmental organizations, Greenpeace focuses on direct action involving drama and excitement. It sought and arranged catchy symbols, framed by activists who put their lives on the line to confront the evil perpetrators. The success of this strategy made Greenpeace a trendsetter in visually exposing environmental conflicts: "Greenpeace positioned itself early on, indeed ahead of the curve, regarding the development of the political and communicative culture that was only then beginning to emerge."[21]

The perfect "timing of moral intervention" via precisely timed release of images and messages is a "necessary condition for effectively staged confrontations."[22] Waiting for the right moment also requires investigation and planning, which set Greenpeace apart from other environmental organizations. Going through the group's documentation on the Brent Spar conflict, it is evident that Greenpeace employees used information-gathering and -processing networks similar to those used in investigative journalism. In the Greenpeace report "Brent Spar or the Future of the Oceans," Jochen Vorfelder notes that a colleague at Greenpeace received a hint about the imminent disposal from a "contact person in the offshore industry." The contact also passed on a report from the University of Aberdeen, which recommended the option of horizontal dismantling and letting the rig sink into the ocean.[23]

A campaign of this scale and scope hinged on strong logistical support. There was a well-stocked storage facility in Hamburg and, under the leadership of veteran campaigner Harald Zindler, ships could leave for action on short notice. All the while, Greenpeace was working on basic media efforts: "Experts were interviewed and called upon for consultation on the phone; photos, videos and briefing papers were put together; press briefings . . . were written up. Journalists were selected and privately briefed on the action."[24] It was becoming clear that this was far from a David versus Goliath conflict—in light of its organizational efficiency, professionalism, and media relations, Greenpeace is a giant in its own right. For the British social scientist Christopher Rootes, the Brent Spar campaign was the strongest evidence that certain environmental campaigning organizations were sometimes more powerful than their mighty opponents: "It appears that it is environmentalists who have made greater, more creative and more effective use of these new technologies than the corporations they oppose."[25]

On April 30, 1995, Greenpeace International began the occupation. Dutch, British, and German employees jumped into action, but in its initial stage, the campaign was far less successful than originally planned. Due to stormy seas and technical problems, the first crew to land on the rig lacked equipment. The team nevertheless comprised the key players: "Six climbers and activists, a camera team, two photographers and two reporters, with the help of anchors, climbing ropes and a climber's net, had captured Brent Spar."[26] Leading media outlets covered the occupation: the campaign began to make its way into everyday media coverage and to gather momentum as activists remained aboard the rig for three weeks. During this period the Scottish authorities gave final approval to the sinking of the oil rig. All the while, Shell was doing everything it could to expel the occupiers through legal means. The first eviction order was served by a Scottish court and had no effect. The crew on the rig changed every week: "New journalists came with the new radio crew, activists and sailing team. They reported the everyday life of the occupation and, frustrated by the day-to-day routine, they conjured up the image of a romantic life on the waves,"[27] as Jochen Vorfelder described the media's role during the first few weeks of the campaign. The journalists became part of the action. It was a form of "embedded journalism."[28]

FIGURE 6.1. Enter Greenpeace: activists approaching the Brent Spar in June 1995. Copyright © Dave Sims/Greenpeace.

During this period samples were taken, which, some weeks later, would lead to false data on the level of toxins on the rig. Although the media did report on the action, this failed to attract public attention. It was the eviction that heralded a breakthrough for the political campaign.[29] Shell sought the eviction notice in a Scottish court and it was served on May 19. Employees of Shell and police arrived at the Brent Spar three days later with a large construction platform and gradually cleared the entire rig in spite of the rough seas. The process took all day because some of the activists had chained themselves to parts of the facility, retreated to the tops of towers, or hid in the depths of the rig. This direct confrontation stimulated media attention. Once again the campaigners demonstrated their skills in dealing with visual media. A freelance British photographer hired by Greenpeace to cover the occupation reported that a member of the Shell team had forbidden him to take pictures and had blocked the camera lens with his hand: "I kept pressing the shutter and said 'Thanks, sir!' And he stared at me and said 'What for?' So I said: 'Sir, you've just given us a great shot. Shell stopping freelance photographers from taking pictures. Do you understand, sir?'"[30]

Over the following days, media attention somewhat declined again, but the issue moved onto the political agenda. At the end of May, the youth organization of the Christian Democratic Party (Junge Union) of North Rhine-Westphalia was the first German organization to call for a boycott. Politicians of different stripes followed suit. Angela Merkel, at the time the German minister of environmental affairs, spoke out against the sinking of the rig and begged for support from her British counterpart, John Gummer, during his visit to Germany, but she failed to leave an impression. She noted afterward that the plan was in line with international law and that the sinking of the rig would be an exception, though she herself would have favored a different method of disposal.[31] The UK government was still firmly committed to Shell's disposal plans and was bracing itself for the upcoming confrontation during the North Sea Conference. Not only the German government voiced disapproval: during the Brent Spar occupation, the European Union's environmental commissioner, Ritt Bjerregard, and the Danish environmental minister, Svend Auken, spoke out against the plan. In fact, Sweden, Denmark, the Netherlands, Germany, and Belgium had all collaborated in Esbjerg to launch a European motion to ban the sinking of offshore oil rigs, but vetoes from Britain and Norway blocked the initiative. While Norway had generally opted for disposal of its facilities on land, it did not want to ban the sinking of rigs outright.[32] In the end, the conference passed a resolution calling for disposal on land, and everyone but Britain and Norway signed. Some observers suggest that governments were only too happy to focus on the disposal problem, as it drew attention away from other, arguably more pressing topics such as chronic pollution from other sources and overfishing, where responsibility was more evenly shared. After all, it was easy for countries with few or no oil fields to call for disposal on land.[33]

The protest was gathering momentum, and in Germany more than anywhere else. At first glance, this was not a big surprise: British authors noted that the rig was basically no different from other scuttled ships.[34] But in the weeks following the occupation, Germany witnessed the launch of what would become one of the biggest consumer boycotts in postwar history. There were several reasons for the huge appeal of the boycott. First, there was the visual appeal of the campaign's images. Another reason was that the rugged northern North Sea had potential as a canvas for individual adventure fantasies. The weather-beaten rig—rusty, bleak, and desolate—no longer served as a symbol of technical progress. It was more a shabby relic of bygone times. In contrast, the Greenpeace activists who mounted the rig were agile, courageous, determined, and passionate. Their physical presence in this austere environment seemed to evoke a modern form of piracy for a good cause. Who would put themselves at the mercy of such an extreme landscape just for the fun of it? The public began to take a genuine interest.

The occupiers' image contrasted sharply with the anonymity of the corporate giant Shell. As it happened, the German branch of Shell had initiated an image crusade in the run-up to the Brent Spar campaign, which provided a perfect launching pad. In an effort to address Shell's lack of a distinct corporate profile, the firm had kicked off a new advertising campaign in Germany on March 1, 1995. The slogan was, "We care about more than cars." Since Shell was selling fuel that looked no different from the oil of other companies, it sought to create a brand that "represented not only the products, but the entire company."[35] In this way, the anonymous product turned into part of a framework of social responsibility and thus gave Shell a unique selling point on the fuel market. A new image would then earn the company more sympathy among consumers and a competitive edge. Social issues were at the core of the campaign: road rage, discrimination against disabled people, and prevention of waste. The latter issue kept coming up in connection with Brent Spar. From the middle of March, one-page ads were placed in newspapers, alongside the insistent slogan, "We'll change *this!*"—a gift for the activists' campaign strategy. Thanks to the proposed sinking of Brent Spar, the activists were able to call out this contradiction and, with elegant impudence, skim off the cream of Shell's advertising campaign, and use the heightened attention of consumers for their own ends. In parallel with the occupation of the rig, Greenpeace occupied Shell's advertising and used the company's logo, tainted with oil and including a mock dialogue:

> Shell: "We care about more than just cars. We dump oil rigs into the sea, pollute the environment, poison habitats, threaten the eco-system, and ignore alternatives."
> Greenpeace: "We'll change this!"[36]

In short, Greenpeace turned Shell's quest for a caring image around and depicted the company as a careless, ignorant leviathan, which touched a nerve in the wider public.

On June 1, Greenpeace Germany released the results of a survey it had commissioned from the Emnid opinion research institute. According to its figures, 74 percent of Germany's citizens were prepared to boycott Shell gas stations. Since the eviction of the activists, Shell's UK branch had conducted public relations work on the environmental effects. Nonetheless, there was little public awareness about its educational efforts. Shell also sought to avoid further confrontation with Greenpeace. During the night of June 11–12, when five tugboats began to move Brent Spar from the northern North Sea toward the disposal area in the Atlantic, two Greenpeace boats followed: the organization was unrelenting. Greenpeace made it known that it would do everything possible to prevent the sinking of the rig.[37] The confrontation over Brent Spar further inflamed the boycott. Shell Germany, though not directly responsible for what happened in the North Atlantic, found itself wrestling with a growing mass of indignant consumers. Every TV channel provided in-depth coverage of demonstrations at gas stations. Consumers seemed united in their goal of a full boycott. The mission brought together all sorts of people: actors, church leaders, government ministers, consumer organizations, politicians from the Left to the far Right. Even the European Car Club and the advisory council of Shell Germany came on board. The broad consensus was that Shell should refrain from its plan to sink the Brent Spar.

Shell caved in. At 7:00 in the evening on June 20, 1995, the corporation announced its decision not to sink the oil rig. It was not Shell UK that had changed its mind—it was the Dutch head company, the Royal Dutch/Shell group, that resolved the matter. It emphasized that it was unshaken in its opinion about the deepwater disposal method as the most environmentally friendly option. But the company yielded to public pressure, rather than Greenpeace's ongoing occupation of the rig. Chris Fay, the chief executive of Shell UK, blamed the environmental agencies of Europe, who had decided to undercut long-standing agreements on the occasion. He announced that he would apply to the British authorities to dispose of the rig on land. Prime Minister John Major was *not amused.* Just two hours earlier, he had again emphatically restated the British government's support for Shell UK's disposal plans during Question Time in the House of Commons. Both the government and the company had been made to look foolish: Greenpeace, consumers, and the public were jubilant. Yet in the weeks that followed, this celebration gave way to a more sober mood as it emerged that the campaign would not, after all, go down in the history of the environmental movement as the perfect success story.

NUMBERS IN PLAY

The trigger for the reassessment was the figures that Greenpeace had publicized in the middle of June. Activists had taken samples during the first occupation and examined them in Greenpeace's own laboratory in Exeter. Greenpeace suggested that much more toxic material might be on board Brent Spar than the 130 tons claimed by Shell. In a press release from the German branch of Greenpeace on June 18, 1995, the activists declared: "While the sampling technique was not scientifically precise, these samples and overall calculations nonetheless support the claim that 5,500 tons of oil and residual oil are still on the rig."[38] Three days later, Shell announced that it would not sink the rig. Following this declaration, the company hired the independent organization Det Norske Veritas (DNV) to conduct an inventory on the rig.[39] Greenpeace had to concede in due course that the results of its sampling had been false.[40] While occupying the rig, they hung a "peanut butter jar and, to weight it down, an iron nut on a nylon cord that was connected to a roll of cables," which, instead of sinking into the tanks, was held in the ventilation pipes, according to Reiner Luyken, a journalist writing in the renowned German weekly *Die Zeit*.[41] The finding called for an apology: "I apologise to you and your colleagues," Lord Melchett, the head of Greenpeace UK wrote to Christopher Fay, head of Shell UK. In this open letter, he admitted mistakes, but also contended that the figures had not influenced the protests themselves.[42] The confession caused a commotion in the German media. Before too long, the faulty calculations became a lie, an intentional misrepresentation. On September 8, 1995, *Die Zeit* ran the headline: "'Sorry': A War over Nothing, the Victory of a Mistake?"

It is rewarding to recap Greenpeace's line of argument. The organization argued from the beginning that the Brent Spar was holding toxic substances that would be released as a result of sinking the oil rig, and that the event would set a precedent for the future. The amount at stake was 130 tons of different harmful substances. These amounts were calculated on the basis of Shell's own data, which it declared to be clear but rough estimates, and not precise measurements. As Greenpeace had justifiably suspected, these amounts could be much higher. To begin with, they challenged the figure of 130 tons. It was not acceptable to deposit any chemical substances illegally in the open sea. The activists at Greenpeace also argued that sinking the Brent Spar would serve as a precedent: over 400 offshore facilities would have to be eliminated in the years to come. It was thus imperative to stop the sinking of Brent Spar, to push oil companies toward an ecologically responsible attitude regarding their facilities, and to stimulate adequate legislation for the disposal of facilities. The environmental organization tried to depict Brent Spar as a mere example of the general problem of pollution of the sea. Even if 130 tons of toxic substances could be absorbed into ecosystems, given widespread pollution and overfishing of the oceans, the extra burden was simply inacceptable. One sunken rig might be permissible—an authorized dumping ground for all offshore equipment was not.

As they had done regarding the false data, Greenpeace offered conclusive proof within only a few days that the incorrect statistics had not caused the boycott, the broad political consensus over the sinking of Brent Spar, or the support of other European environmental ministries. It also pointed to the relatively scant reporting on the figure.[43] Nonetheless, as news of the false figures sank in, media pundits started "for the most part, to rewrite the story of Brent Spar. The hour of the commentators had come."[44]

Both Shell and Greenpeace were struggling for their reputations in the aftermath of the clash. Shell Germany opted for a campaign of apology. A few days after announcing its decision to dispose of Brent Spar on land, Shell placed a huge and largely text-based advertisement, which once more used the slogan that Greenpeace had so effectively used against them. "We'll change *this*" became "*We'll* change this."[45] Peter Duncan, the chair of the managing board of Shell Germany conducted serious conversations and owned up to mistakes, which, as he saw it, were more or less due to the company's poor communications. As the ad declared, "For a decision to adhere to international law was itself not enough—it also needs society's acceptance."[46] The corporation would subsequently ensure a higher level of transparency in disposing of the oil rig. Public relations experts consider an apology campaign risky, but it looked like a good choice under the circumstances: "Beginning at a certain point, the company no longer had an opportunity to make its important arguments in support of sinking the oil rig with the hope of any acceptance. After the public apology, however, it was again possible to get these facts through to people."[47] Shell stuck to the script: the sinking of the rig had been a rational, ecologically defensible, and economical strategy. But they were bowing to public pressure and political whim.

Greenpeace's apology to Shell did not directly influence the credibility of the nongovernmental organization (NGO). While German media commentators were quick to doubt and express concern about the "eco-giant," the general public was unmoved. In fact, Greenpeace's admission reflected well on them and came across as a mark of their integrity. An Emnid opinion survey for the German weekly *Der Spiegel* found that Greenpeace still enjoyed high levels of trust and sympathy. Attesting to this trust bonus was the fact that 28 percent of respondents would definitely vote for Greenpeace if it were a political party, another 33 percent would consider voting for its members, and 70 percent said they would be in favor of giving the organization the Nobel Peace Prize.[48]

However, the mass mobilization in the wake of the Brent Spar campaign would not last. Greenpeace launched a campaign against French nuclear tests on the Mururoa atoll in June 1995 that came to naught. Pointing to this campaign, Greenpeace employee Thomas Schultz-Jagow felt that some of his colleagues had succumbed to megalomania. Enthused by the Brent Spar victory, people inside and outside the organization believed that they were living in a new world where Greenpeace, after taking down Shell, could now do the same with the French government.[49] At the end of 1996, a study assessing the results of the Brent Spar campaign noted that the media had perceived the

Mururoa action as a defeat for Greenpeace. A French navy vessel seized the organization's ship, and the nuclear tests went on as planned. Nuclear tests were arguably a far more important issue, and yet protest failed to stimulate much response. The Mururoa campaign "left behind far fewer traces" in Germany's media than Brent Spar.[50]

The same held true for Greenpeace's action in Nigeria. For some time, human rights organizations had been pointing to Shell's role in the ecological devastation in the Niger delta where oil companies were drilling, and NGOs had sought to draw attention to the plight and political oppression of the indigenous Ogoni tribe. In November 1995, the situation in this region came to the attention of the German media. This was largely due to the imprisonment and subsequent execution of the Ogoni environmental activist Ken Saro-Wiwa, and was also the result of Shell's prominence in the wake of the Brent Spar campaign. Once again, Greenpeace was accused of being too hesitant, particularly compared to its action in the Brent Spar campaign.[51] Moreover, the public was far less engaged than they had been in the campaign around Brent Spar: "A stark sign of this disparity was the letters Shell received in protest: according to the PR manager K.W. Lott, Brent Spar triggered around 16,000 letters; Nigeria only 400."[52]

Subsequent events showed the limits of Greenpeace's campaigning effort: the organization did not retain exclusive control over social protest and mobilization, notwithstanding suggestions from the media to the contrary. It did continue to produce attractive images, but these were not published. In other words, Greenpeace not only depended on the media's responsiveness but also faced competition from other news. The world had not changed overnight and Greenpeace was not emerging as the environmental policeman of the world. It gradually emerged that the North Sea campaign was probably not worth so much attention. Sinking an oil rig in the North Sea seemed harmless compared to nuclear testing and the ecological destruction and oppressive politics in Nigeria.

In retrospect, the gap between the public spectacle and the ecological effect seems to be a far more serious issue than the stir over figures. After all, Greenpeace was an important player in the emerging global environmental policy. In the immediate aftermath of Brent Spar, conventional wisdom held that the campaign defined a turning point for the interplay of civil society and environmental policy. On June 23, 1995, the *Deutsches Sonntagsblatt* declared that the protest "revealed a new level of political culture": the precedent showed that "civil society's resistance can form a majority and be successful."[53] For the sociologist Ulrich Beck, the events showed "the subpoliticization of world society." In Beck's opinion, Greenpeace embraced "direct" politics, which meant "selective individual participation in political decisions, skirting around political decisions and existing democratic institutions (parties, parliaments), often without legal protection." He found a new type of politics, namely, "an alliance between nonparliamentary and parliamentary powers, citizens and governments around the world for a legitimate cause in the highest sense of the word: saving the environment, and thus the world."[54]

Other observers took a more skeptical view. For example, the British historian Eric Hobsbawm argued that Greenpeace had opted for an inferior technological solution. Hobsbawm nevertheless perceived the concern as justified and helpful: "The fact that a world economic power can be forced to pull back on an environmental issue is an important step on the road toward an age in which polluters account for the costs of pollution from the start."[55] But Hobsbawm warned that the success of such a campaign relied not so much on its political or intellectual justification as on manipulative "mass psychology." Hobsbawm found striking parallels with the conduct of media moguls "such as Berlusconi or Time Warner, when they promote a huge film, namely, by addressing the public directly through modern media."[56] In an editorial in the alternative-left newspaper *die tageszeitung* of July 8, 1995, Nicola Liebert framed her skepticism in a rhetorical question: "Who will save the floodplains?" She argued that the everyday work of environmental protection was ill-suited for media events of the Brent Spar variety and thus faded from public view. She made a point that would later resonate in the literature: "A sale of indulgences takes the place of environmental commitments."[57]

These comments reflect a sense of unease about the Brent Spar campaign. Growing professionalization combined with a notorious lack of transparency of its internal structure gave Greenpeace the air of a behemoth that could strike out of the blue, had bypassed institutional structures, and was accountable only to itself. When Greenpeace celebrated its twentieth birthday in 1991, media reports in Germany, the United States, and Australia were surprisingly critical, as journalists raised questions about its hierarchical structures and opaque finances.[58] After Brent Spar, the environmental hero was an even more ambiguous figure, and journalists began to add critical subtexts to articles on Greenpeace. The end of its long honeymoon was a source of irritation for Greenpeace, but it was perfectly reasonable. A transnational environmental organization with a multimillion-dollar budget cannot afford to hide its strategy and its finances from the world. But at the same time, Greenpeace was never secretive about its hierarchies, which it argued were crucial for flexible and effective action.[59]

Another discussion point was the organization's degree of professionalization. While activists in nature conservation and citizen initiatives were often ridiculed for working only locally and knowing nothing about effective communication and strategic positioning, there was also considerable skepticism about the more cunning approaches at Greenpeace. To this day, carefully staged and choreographed events define public relations at Greenpeace. Other environmental NGOs argued that Greenpeace's stunts were stealing the show and diverting attention from other like-minded groups. Furthermore, its events were sensationalist and created inflated expectations.

But while scoffing about Greenpeace, more traditional organizations learned a lot from their rival.[60] At other large environmental NGOs in Germany, membership figures and donations grew dramatically during the 1990s.[61] Critics warned of organizational monoculture, but diversity continued to rule in the institutional landscape. German environmentalism retained a range of organizations with different styles of activism: environmentally minded people have the choice. Having said that, there can be little doubt that Greenpeace receives the largest amount of donations, and many of these donations do in fact have an air of indulgences for sale. Supporters pay so the activists can act in their places, using approaches they are unable or unwilling to carry out. Is this more than a case of a useful division of labor? It certainly is a financial necessity. If Greenpeace does not campaign convincingly and effectively, donations decrease notably. At the same time, proper accounting is a matter of legal obligation, as irresponsible behavior would jeopardize the group's nonprofit status.

As these examples illustrate, ambiguity ran through the comments in the wake of the Brent Spar campaign. Few observers challenged Greenpeace as a matter of principle: they widely agreed that its work was important for society. And yet it seemed that the campaign did not have the right balance between material danger and public excitement. *Der Spiegel* ran a cover story in the aftermath of the campaign that dubbed the boycott "the hypocrites' protest."[62] Even twenty years later, it is not clear what we should make of the campaign.

Brent Spar was dismantled in an orderly fashion. As it happened, some parts of the facility did eventually end up in the sea, and became part of the foundation in a Norwegian harbor. Hazardous substances were disposed of according to the guidelines. And Brent Spar set a precedent: the proper method for disposing of facilities of its size is now codified in law. The deconstruction and recycling of the large offshore structures of the oil and gas industry have now become a lucrative line of work.[63] But these operations have also begun to trigger protests from nature protection organizations: in the North Sea and the Baltic Sea, populations of starfish and mussel banks have flourished on the large concrete plinths of the drilling stations that are awaiting disposal, and these populations provide fish with protection and act as nurseries.[64] But exceptions are no longer in the cards: starfish are picked off the pillars, muscles are scraped away, and the building material goes back on land. In a way, Brent Spar highlighted the difference between grandiose plans and realities on the ground. Environmental priorities are always open to debate: who defines the value of clean water, a bed of muscles, or a human life that becomes endangered such as during the dismantling of Brent Spar? Moral dilemmas have many dimensions.

When Shell caved in at the end of June and decided to dispose of Brent Spar on land, there was much celebration and also a period of harsh recriminations. Greenpeace quickly had to confront a deep-seated challenge to its status as a credible source of environmental information. The politicians were accused of opportunism when subsequent environmental policies showed that Brent Spar had not heralded a more progressive and consistent green agenda. The media went through a period of self-reflection. The boycotters were now no longer needed—and returned to Shell's gas pumps.

As it happened, one of the most important long-term effects of Brent Spar was that it inspired critical thoughts about the group's different roles. Greenpeace was particularly thorough in this regard. As an NGO relying on donations, Greenpeace depended on high marks for credibility. They could not afford to lose public confidence.[65] The flawed figures underscored concerns about the lack of transparency. Greenpeace Germany at least did some soul-searching, as recalled in the book *Greenpeace*, which was published on the German branch's twenty-fifth anniversary.[66] It offered critical reflections on media strategies or the power of pictures and campaigning. However, the volume also offered classic success stories, mirroring the philosophy of Greenpeace as an amalgam of different principles. The organization also published a comprehensive account of the Brent Spar affair in 1997. It was a nuanced and balanced presentation that recognized ambiguities and reflected the group's move toward greater transparency. A supposedly erratic organization was now opening itself up to a plurality of opinions.[67]

In a wider context, Brent Spar raised doubts about the integrity of environmental activism. It drew legitimacy from its claim to being "good" and being "right." But moral arguments are always open to debate. Journalists gleefully nourished stereotypes of whining do-gooders. From this perspective, Brent Spar "marked a watershed moment for the credibility of environmental protection campaigners."[68] Greenpeace also had to address accusations of "knowingly misleading the public."[69] The campaign was filed in the folder of "media fairy tales," purposefully cultivated by Greenpeace itself: "The goal of the campaign was a media spectacle based on pure self-interest."[70] The journalist Michael Kröher blamed environmentalists for their "apocalyptic worldview," and considered the Brent Spar campaign nothing but "a catastrophe."[71] These authors portrayed Greenpeace as a greedy protest machine, while in the same breath decrying the naivety of the media and consumers who were taken in by them. Commentators reacted angrily to the supposed moralistic untouchability of Greenpeace.[72] The Brent Spar episode opened a door to challenging the integrity of environmental campaigning and underscoring negative stereotypes. The symbol had assumed an existence separate from the bigger story for which Greenpeace had staged it in the first place.

THE MERITS OF CAMPAIGNING

So how does one measure the value and the effects of the Brent Spar campaign? It achieved its immediate goal—a ban on sinking offshore rigs into the depths of the oceans. In 1998, this became part of the *Convention* for the Protection of the Marine Environment of the Northeast Atlantic (OSPAR), a mechanism by which fifteen governments and the European Union cooperate to protect the marine environment of the Northeast Atlantic. It is reasonable to assume that this would not have been achieved without a powerful symbol and public attention. But was it worth all the outrage? Scientific experts offer different opinions on the environmental effect of sinking oil rigs. While some are convinced that the effect would be minimal and that the ordinary seepage from deep sea oil sources is far more serious, other scientists dismiss the possibility of making these assessments, given how little we know about the seabed. The British authors Tony Rice and Paula Owen have provided

the most balanced account of the different arguments. They accept Greenpeace's criticism of the figures Shell gave to the British authorities during the
approval process, considering that, initially, it was not possible to know for
sure which toxic elements would be discovered on board the oil rig. Only
after the event—following the assessment by the independent Norwegian
firm DNV—was the information everyone required at the beginning made
available.[73] They even acknowledged Greenpeace's argument that it would be
hard to forecast the effects at the site because not enough was known about
the biological conditions in this region. Ultimately, Greenpeace was right
to assert that, even assuming the damage at the site would have only a limited effect, we could not be sure that the rig would not break up while being
sunk. This would have the effect of scattering the substances over a far larger
area. In their conclusion, Rice and Owen imagine the scenario without the
Greenpeace activists: the rig would have been sunk, the local sediment a little
disturbed, a few animals would have died, and the 130 tons of oil, PCB, and
other materials would have slowly bled out of the wreckage and become part
of the cycles of nature. In short, they argue that from a scientific point of view
(Rice's field is marine biology), it is likely that the sinking would not have had
a great effect. However, they emphasized that this did not mean we should
now vote in favor of sinking oil rigs.

Furthermore, Rice and Owen maintain that Greenpeace was justified in
assuming that sinking this oil rig would have set a problematic precedent.
However, in their opinion, the most harmful consequence of such a scenario would have been the continued use of this disposal method without
further investigation—and without the involvement of other stakeholders. In
short, there would have been little critical dialogue.[74] After all, Greenpeace
had lifted the veil of secrecy over the negotiations between state authorities
and industry.[75] From a scientific perspective, Greenpeace had exaggerated the
possible ecological effects of sinking Brent Spar.[76] However, from an ethical
standpoint, their argument was strong.

Twenty years later, it is clear that the Brent Spar campaign has left a powerful and permanent mark in the iconography of the environmental movement. A distinctive culture of commemoration has grown around the event.
When Greenpeace Germany planned to mark the ten-year anniversary of the
Brent Spar campaign in 2005, its friends from Greenpeace UK were skeptical
and pointed to the flawed numbers: "Stop digging when you are in a hole!"[77]
Greenpeace Germany saw things differently: the organization compiled a

commemorative report that did not spare the data issue and the brief false alarm.[78] It was an exercise in self-praise and self-reflection. The opportunities and limitations of the campaign were explored once again. The conclusion was that the promise of the campaign had not held true: "The movement that built up around Brent Spar did not usher in a new era of consumer power, but rather proved to be—up until now—its peak."[79] The statement matched a culture of commemoration that routinely includes a critical slant, but it was also clever marketing: Greenpeace Germany sold the data issue as a call for more transparency. The organization recognized that it was in a precarious position: doubts about the credibility of Greenpeace put donations and support in jeopardy. As one of the consequences of the Brent Spar affair, Greenpeace Germany created its own research department for the verification of factual statements and data before publication. It also established double-checking (henceforth, they would always take two samples for testing) in order to lower "the risk of reputation damage from false measurements."[80]

On the other hand, Brent Spar became an example of how an embattled industry could make matters worse through a disastrous public relations strategy.[81] The Shell corporation changed its strategy on corporate responsibility and crisis management.[82] The company sought to avoid similar confrontations in the future and embarked on proactive green communication. In other words, one of the campaign's results was a greenwashing effort. Confrontational NGOs should not again be in a position to sift through the company's dirty laundry.

The narrative of Greenpeace is upbeat. In the spring of 2015, Greenpeace celebrated the twentieth anniversary of Brent Spar with massive publicity: video clips, historical image series, press releases, and eyewitness accounts.[83] Part of the effort was a scholarly assessment and comprehensive review of disposal practices in offshore industry.[84] Brent Spar serves as an enduring icon for Greenpeace, reassuring the organization and its supporters of the importance of resistance and action. Twenty years on, Greenpeace Germany has not staged a campaign on a similar par.

7

A LANDSCAPE OF MULTIPLE EMERGENCIES

NARRATIVES OF THE DAL LAKE IN KASHMIR

SHALINI PANJABI

THE IMAGE THAT KASHMIR evokes in the wider imagination is of a beautiful valley trampled in recent decades by unrest and violence in the conflict over sovereignty. The conflict and its complex politics have had a major effect on Kashmir's ecology, on its celebrated landscapes, including the iconic Dal Lake. Contemporary descriptions of the lake, in books and the media, commonly refer to it as "the crown jewel of Kashmir." The Dal Lake is one of the few inhabited lakes in the world, with approximately 50,000 people living within it. They live in small hamlets on islands and in houseboats. A large number of people also live around its shores. Most of the vegetables for the city are grown in the lake and it is a source of drinking water too. It is also a space of leisure for the city's residents, and critically the emblematic center of tourism in the valley. Yet the lake is now described as "dying"—with its area shrinking and pollution levels rising. In the past fifteen years, after the violence in the region abated, various agencies have turned their attention to "reviving" the lake. However, the various plans and actions, all imbued with a sense of great urgency to "save" the lake, have achieved little. Multiple interlocking and often persistent emergencies have arisen, yet approaching the problem as an environmental apocalypse has not been productive. The straight narrative of an impending catastrophe has led away from a comprehension of the myriad complexities, contradictions, and transactions therein.

The Dal Lake (or the Dal, as its called in Kashmir), is located in Srinagar, the historical, economic, and political center of Kashmir. Kashmir is commonly understood as referring to the Kashmir valley—often called just "the Valley" in India.[1] The valley located in the western Himalayas between the Himadri and Pir Panjal Ranges is around 140 kilometers long and 40 kilometers wide. Kashmir today is divided, with most of it under India's control, and a small area under Pakistan's control. I will briefly recapitulate the history of the conflict to give a sense of the contested environment and of some of the forces at play.

THE KASHMIR CONFLICT

The modern state of Jammu and Kashmir was created in 1846 through a treaty signed by the British East India Company and the Hindu rulers of Jammu, the Dogras.[2] At Independence, the Princely States in India were free, at least officially, to join either India or Pakistan or remain independent. Maharaja Hari Singh, the Hindu ruler of the Muslim majority State of Jammu and Kashmir, dithered in his decision probably in an attempt to remain independent. Rebellion was also brewing in certain areas against his unpopular rule, when armed "tribesmen" from Pakistan moved into the valley in October 1947. Hari Singh requested the help of the Indian troops to fight the intruders, and the Indian government agreed on the condition that Hari Singh should accede to India. Hari Singh signed the Instrument of Accession and overnight Indian troops were sent to Kashmir and the tribal militia pushed back. However, regular skirmishes between Pakistani and Indian troops began soon after and have continued to this day—with major escalations in between.

Hari Singh's decision to accede to India was never fully accepted by both Pakistan and many Kashmiris, and this forms the kernel of the conflict. Pakistan has maintained that Kashmir, as a Muslim majority state, should rightly have joined Pakistan, while Kashmiris have also nurtured the hope of forming an independent state. With Pakistan-supported guerrilla groups becoming increasingly active, and India continually reneging on even its limited offers of autonomy, protests against Indian rule grew in the ensuing decades. Elections to the State Assembly in 1987, widely perceived as being heavily rigged, triggered widespread protests that turned increasingly violent. The government's response was heavy-handed, and force was disproportionately employed. The situation spiraled out of control and the first half of the 1990s was violent and traumatic for Kashmiris. The period of intense violence receded when the Indian government began to accept that force was not the answer, and gradually talks and negotiations were initiated. The power that the militants wielded also corrupted many, and they became preoccupied with kidnappings and extortions and various illegal trades. By the mid-1990s, most Kashmiris were exhausted by the spiral of brutal violence and anomie in the valley, and support for the militancy began to wane.

Tragically, although the Kashmiris have endured enormous suffering through this period, they have achieved no political gains. The Indian government has taken barely any political initiatives and continues to rely heavily on the army to maintain its control in Kashmir along with special economic and social "packages." However, the precarious peace thus achieved can break easily, as happened in 2016 when the valley was paralyzed for over four months, following protests over the killing of a young militant by the police. Yet even as the valley remains volatile and the Kashmiris alienated, the regional and central governments are eager to herald a return to "normalcy."[3] The Dal Lake remains a centerpiece of these attempts, with photographs of visiting tourists used to indicate that peace has returned to the valley.

KASHMIR IN THE WIDER IMAGINATION

The position of the Indian government on Kashmir is largely backed by the Indian public. Opposing the movement of any part of India to secede, they seek to defend the country's "territorial integrity," but in Kashmir's case, emotions are heightened because of the hold it has on the Indian imagination.[4] For Indians, Kashmir has been a land of incomparable beauty with its high mountains crisscrossed by rivers and lakes and flowering meadows. Its temperate climate also makes it the land of escape from the summer heat of the Indian plains. Immortalized in Hindi movies and song sequences, it has been the dream destination for Indian tourists, especially before the possibilities of travel abroad opened up. Adding to the allure of Kashmir are its horticultural and agricultural produce, including a high demand for its prized saffron, almonds, walnuts, and apples. The rich handicrafts—woven carpets, embroidered shawls, and carved wooden products—have also always been highly valued by wealthier Indians and Europeans.[5]

The inscribing of Kashmir as an earthly paradise owes much to the Mughals. In their imagining, the whole valley itself was an idealized Persian garden.[6] Kashmir was part of the Mughal Empire from the end of the sixteenth century to the early eighteenth century. Akbar was the first Mughal emperor to visit Kashmir in 1589, and it was his son and successor Jahangir along with his wife, Noor Jehan, who popularized the idea of Kashmir as a summer retreat.[7] With a large retinue of courtiers and officials, they spent many summers there, away from the dust of the Indian plains. Jahangir and his successor Shah Jahan also had a huge influence on the architectural fabric of Kashmir, building many celebrated gardens, mosques, and palaces in the valley. More than seven hundred gardens were built through the seventeenth century.[8] These gardens were distinguished by their terraced waterways, pavilions, and avenues of *Chinar* (Oriental Plane) trees. They largely followed the contours of the hilly terrain, with water from nearby springs or streams directed to the water channels and fountains. After traversing the gardens, the channels would usually flow into a nearby water body. Three of the largest and most well-known gardens—Shalimar, Nishat, and Chashme Shahi—were set around the shores of the Dal Lake.

In the nineteenth century, the British interest in Kashmir was fueled by travelogues and fictional writing. A prominent example was Thomas Moore's hugely popular narrative poem *Lalla Rookh*, published in 1817. It depicted the romance between a daughter of the Mughal emperor Aurungzeb and the king of Bukhara, through travels in Kashmir. The poem waxed eloquent about the singular beauty of the "vale of Cashmere," "the most lovely country under the sun." The Shalimar Gardens and the Dal Lake were exotic backdrops for romantic longing:

> *Who in the moonlight and music thus sweetly may glide*
> *O'er the Lake of Cashmere, with that One by his side!*[9]

Moore's rhapsodic vision was particularly ironic because he had never traveled to Kashmir, and his idealized images of the valley stemmed from his reading of travelogues, primary among which was François Bernier's *Travels in Mughal India*.[10] The success of *Lalla Rookh* marked the beginning of the British enchantment with Kashmir, and motivated travelers to visit the valley themselves. Among these tourists was landscape photographer Samuel Bourne, whose "spectacular" and "picturesque" portraits helped create the template for depictions of the mountainous landscape.[11] Kashmir soon emerged as the favored summer destination for British officials and others who camped in the temperate climates with their families and retinues. The houseboats of Dal Lake are popularly touted as the British contribution to the Kashmiri landscape.[12]

After the tumult of Independence and partition had settled, from the 1960s onward, tourists from across India began visiting Kashmir. They were now the main customers of the houseboats, as Dal Lake remained the focal point and tourism thrived in the valley. Change came abruptly in 1989, as the conflict turned increasingly violent and normal life was disrupted. For the next decade, barely any tourists visited—except Hindu pilgrims on an annual pilgrimage in the valley. A different set of images about Kashmir now circulated in India, resonant with the anger of Kashmiris. The response of the Indian state and the public though was unyielding—Kashmir was an integral part of India and would remain so.[13]

As the violence ebbed and the administration regained control, tourists gradually began returning to the valley. With traffic back on the streets, shops reopening, and tourists milling around the Dal Lake, life in the valley seemed "normal" again. Keen to maintain a semblance of normality and to help revive the economy, the state and central governments have urged tourists to return to "paradise." Questions raised about the unresolved conflict are answered using images of a busy Dal, proclaiming business as usual.

Dal Lake has also simultaneously been the focus for Kashmiris over the past two decades, with fears being expressed over its impending "extinction." Described as "the timeless mascot of the valley's beauty," it is also the source of livelihood for thousands of Kashmiris. With the lake's waters becoming increasingly polluted and its area seemingly shrinking, divergent narratives have emerged over its "revival."

LOCATING DAL LAKE

Srinagar, the capital of Kashmir, is a fast expanding city with a population of about 1.2 million. It sits at the base of a broad valley at an altitude of 5,500 feet. The rapid unplanned urbanization it has undergone, with inadequate infrastructure, makes its chaotic cityscape similar to other Indian cities. Like many of them it also has an older section with more vernacular architecture, set among narrow, winding lanes, while the newer section is more spread out with largely undistinguished structures of concrete. What sets Srinagar apart though is the large presence of security forces on the streets, and the more enduring presence of water bodies throughout the city. It is a city oriented around water, primarily located along the banks of the river Jhelum. The Dal Lake is part of a hydrological network, along with the adjoining lakes of Anchar, Gil Sar, and Khushal Sar, which are all interconnected with the Jhelum. Canals and channels crisscross the city, and while many of these are now clogged, some are still used for navigation.

The Dal is a very old lake but exactly how old is unclear because its geological and social history are both hazy. Some attribute its origin to the shrinking of an Ice Age glacial lake, but it is more widely accepted as being an oxbow lake. At present, the lake gets most of its water from glacial melt and streams flowing through a catchment area of more than three hundred square kilometers. This water enters the lake primarily via a perennial inflow channel on its northeast side, the Telbal Nallah. Springs rising from the lakebed contribute about 10 percent of the volume of water. The Dal is a shallow lake with four main interconnected basins: Gagribal, Lokut Dal, Bod Dal, and Nagin.

In popular belief, the lake has mythical origins and finds mention in ancient Sanskrit texts. Slightly more definitive citations are quoted from the twelfth century historical chronicle, *Rajatarangini* of Kalhana. Some bunds and channels around the lake are said to date from the fourteenth century. One notable example is the Nallah Mar, a now demolished canal that linked the Dal to Anchar Lake, which is widely believed to have been constructed during the reign of Sultan Zain-al-Abidin in the fifteenth century. More documentation is available from the time of the Mughals, including passages extolling the lake's beauty in Persian texts and poetry composed in the Mughal courts. On their visits to Kashmir, the Mughal emperors and officials chose to live largely around Dal Lake, building citadels and palaces like Hari Parbat and Pari Mahal. As noted, the three most well-known Mughal gardens—Shalimar, Nishat, and Chashme Shahi—were also developed around the lake.

The attraction of the lake for rulers has persisted over time. The kings of the Dogra dynasty in the nineteenth and twentieth centuries also built their palatial residences around the lake. The residences of the chief minister, governor, and other high officials in postindependent India were also planned in an area very close to the lake—and are still there. A fifteen-kilometer-long road, called the Boulevard, skirts the southwestern edge of the lake. It was built during Dogra rule in 1931, by filling up marshy land around the lake. The beginning of the Boulevard is barely a couple of kilometers from the city center and forms a hub for tourists. At the top of a hill across the Boulevard is the Shankarcharya

FIGURES 7.1 AND **7.2.** Daily life on and the vegetable gardens of Dal Lake.

temple, a mandatory stop on any tourist itinerary. This earlier section of the Boulevard accommodates mainly hotels, restaurants, and shops alongside a few government buildings. Some of the hotels are luxury hotels, converted residences of the Dogras, while others are newer, more modest constructions. A golf course also nestles in the foothills. Further along the road are up-market residential colonies, built on reclaimed marshy land, which over time has supplanted the margins of the lake. Interspersed between the hotels and the residential colonies, along the Boulevard, are the three Mughal gardens.

The Dal is the centerpiece of tourism in Kashmir. The image of the lake, with colorful *shikaras* (small boats) and the Zabarwan hills reflected in its calm waters, typifies the vale of Kashmir. It is the most recognizable symbol of the valley, proclaiming its easy beauty while also setting it apart from other Himalayan regions in the country.[14] The first view of the lake from the Boulevard is of a rather busy and crowded waterfront, with people and shikaras milling about, backed by rows of houseboats. The mobility of the houseboats is limited, given their size and style of construction, so they are moored to pillars in the shallow lake and connected to a smaller boat or more usually a small house in which the owners live. As one moves farther along the Boulevard, the traffic on both the road and the lake gradually decreases and a clearer stretch of water is visible.

A short ride inland, however, reveals a different world within the lake. There are hamlets on islets, people living in boats and houses, shops and schools, footbridges, narrow channels between floating gardens, and residents moving around on boats. It is an amphibious world that fascinates many, but is now threatened by calls to "clean" the lake. The people who live on the lake are deemed to be the biggest threat to its existence and the government seeks to move them.

It is difficult to ascertain the precise numbers of people living on the lake. The 2011 census enumerated a population of 33,288, but the media, voluntary groups, and other government departments estimate a higher number of around 50,000 people, living in 50 hamlets within the lake. Some of them have proprietary rights under the Revenue Settlement of the late nineteenth century; while the documentation is not very clear, presently residents are claiming rights to 300 hectares of agricultural land and more than 670 hectares of water. Most of the lake's inhabitants are from the community of Hanjis, and they cultivate vegetables on the islets and floating gardens—alongside some fishing. Many of the families also own and rent shikaras and are thus engaged in the tourist trade, and some of them also own houseboats. Many of the younger generation though are now looking for employment opportunities outside the lake and taking jobs in other parts of the city.

Most of the vegetable plots in the Dal are located on artificial islands or "floating gardens," created through a process akin to reclamation. Vegetables are grown on rafts of closely packed reeds, which are anchored to the bottom by willow or poplar poles. As the reeds decompose, more layers can be added, and in a few years a floating island results. Over time, soil is added to the island, and thus is land created out of water. The floating islands can be towed and adjoined to dry land, to expand an existing plot. Various vegetables are grown on the islands, including tomatoes, squash, radishes, turnips, eggplant, greens, peas, and melons. Lotus grows abundantly, and apart from its flowers it is also harvested for its stems and seeds, along with water chestnuts. These are all local delicacies and sell well, as does the fish. Rush growing on the lake is cut periodically, dried, and woven into mats. The lake is thus a thriving wetland ecosystem, a fertile space for agriculture. Vegetables are produced in large quantities and meet a sizable share of Srinagar's demand. Even in 1895, Walter Lawrence, the settlement commissioner of Kashmir, had noted that the "cultivation and vegetation of the lake is . . . of great importance to the city people."[15] The current value of the annual production of vegetables is estimated to be around $6 million. The trade in vegetables is also a photographer's delight and many tourists visit the floating vegetable market on the lake. This is a daily market, with no kiosks or shops, active for only about an hour at the crack of dawn. Cultivators arrive on their boats with freshly harvested vegetables and flowers, which they sell to traders who also arrive on their boats. There is a lot of hustle and bustle as produce is selected, prices negotiated, and the vegetables weighed and transferred, as the boats constantly bump into each other. The traders supply the vegetables to other vendors in the city, and some sell directly through their own stalls and shops.

As noted, apart from sale of produce, the other major source of revenue for the lake dwellers is the tourist trade. A shikara ride on the Dal, gliding through the waterways (in the constant company of vendors on smaller boats selling flowers, saffron, shawls, and more), photographing and being photographed, is de rigueur for any visitor to Kashmir. The shikaras do a brisk business when tourists are around, but unfortunately, this has often not been the case over the past two decades. The periods of unrest affect occupancy in houseboats even more. With continuing uncertainty, practically no new houseboats are being constructed because no one is willing to invest the large sum of money required. So most of the houseboats on the lake are more than fifty years old now, and many are in dire need of repair.

The years of violence and anomie in Kashmir have affected both the ecological and social architecture of Dal Lake. With the steep decline in tourist numbers, the inhabitants of the lake have seemingly found land more remunerative than water. From all accounts, more islets and floating gardens were created during these past two decades than earlier. The period of lax governance also emboldened some to construct unauthorized houses, and the population registered a steep increase. The countrywide decennial Census of 1991 could not be conducted in the valley, but the census records for 2001 to 2011 indicate that the population of the lake practically doubled in that decade. The numbers rose from 16,766 to 33,288, with a corresponding increase in the number of households from 2,514 to 5,652. The increased population has led to a greater stress on the lake's resources.

Conflicts also generate their own economy, however, and in Kashmir big benefits have accrued to some sections of the bureaucracy, the political class, and the militants.[16] As the central government sought "to win the hearts and minds" of Kashmiris, funds were generously disbursed under various schemes. These offered opportunities for aggrandizement for some in the valley, and, ironically, Srinagar has witnessed a construction boom over the past two decades. Within the lake only shanties and sheds have come up, but along its shores much larger houses and hotels are visible. As the violence ebbed, investments were made in new businesses—hotels, shops, and restaurants—while older ones were renovated. All of these discharge their effluents into the lake. The untreated sewage from the houseboats and houses on the lake flows into it as well. The volume of pollutants entering the lake has increased during the past two decades, and the unrest has often hampered routine cleanups.

DIVERGENT NARRATIVES AROUND THE LAKE

It is largely accepted that Dal Lake's ecosystem has deteriorated through the period of the militancy and its aftermath. The quality of water has declined with the increased discharge of sewage, solid waste, and other effluents—including pesticide and fertilizer runoff from the watershed. This has led to a substantial influx of nitrogen and phosphorus into the lake, resulting in nutrient enrichment and eutrophication. Moreover, excessive growth of weeds and algae chokes the lake's flora and fauna. Lake fishery has declined, and local species of fish have dwindled.[17] The concurrent degradation of the lake's catchment area has led to increased siltation and sedimentation.

The condition of Dal Lake has caused alarm in Kashmir, though the narrative of the declining lake is not new. At the end of the nineteenth century, Walter Lawrence wrote with concern about the lake "silting up." He further expressed apprehension that "unless great vigilance is shown the floating gardens of the lake will be extended, and the already narrow waterways to the Mughal gardens will become blocked to boat traffic."[18] Since then concerns about the lake have been expressed regularly, and with increasing urgency over the past fifty years. During the 1970s and 1980s, the state government commissioned studies and engaged national and international consultants to suggest measures to clean the lake.[19] The consultants made various proposals, of which some were approved and programs initiated, but little happened on the ground. As the insurgency paralyzed the state administration, this, like other governance issues, took a backseat throughout most of the 1990s. In 1997, the state government set up LAWDA, the Lake and Waterways Development Authority, which has since emerged as the nodal agency for issues concerning the lake.[20] Another important development was the filing of a Public Interest Litigation in 2000 by a Kashmiri law student in the Supreme Court of India, against various authorities—including the state and central governments and LAWDA. The petition claimed that the people of Kashmir had been deprived of a clean and healthy environment, and urged the Court to intervene to save the Dal Lake. The courts thus got involved in the proposed cleanup of the lake, and since then have been regularly monitoring the process, issuing notices, and summoning officials to explain the scant progress. As the political situation has gradually eased, more government funds have been released and LAWDA and the courts too have become increasingly active. The central government provides the major share of the finances for the cleanup, thus highlighting its investment in the lake. Funds have been disbursed by the center under the National Lake Conservation Plan (a programme for the conservation of urban lakes throughout India), and the Prime Minister's Reconstruction Plan (a special economic package for Jammu and Kashmir). More than $150 million has been spent so far under these centrally sponsored schemes, besides the expenditure incurred by the state government.

The attention of environmentalists, voluntary groups, the media, and other citizens of Srinagar has also been focused on the Dal. Billboards in the city exhort the lake to be saved, various individuals and groups have proposed solutions, and the media highlight the issue with regular reports and editorials. Many Kashmiris see Dal Lake as an emblem of Kashmir, and the very existence of the city of Srinagar as linked to the lake's condition. Thus the chief justice of the Jammu and Kashmir High Court proclaimed at one of the hearings, "Srinagar exists because of this lake. If it is not there, there won't be Srinagar."[21] An editorial in a leading daily *Greater Kashmir* also made an urgent appeal for saving the "dying water body," as "Kashmiris may not abandon Srinagar in spite of the worst atrocities and harassment faced by them for [the] last couple of decades but the ultimate transformation of the lake into a stinking marsh may force all the citizens to abandon this 2,000 year old City of the Sun."[22] The heightened urgency to save Kashmir's "living heritage" has led to a slew of proposals being attempted to help "revive" the lake.

Some of these are more technical measures that include intensive removal of weeds, setting up sewage treatment plants, improving the circulation of water by declogging water channels, trapping the sediments in a settling basin, and desilting the peripheral areas through suction dredging. Proposals of anti-erosion measures for the catchment zone include the control of grazing, building of check dams, and afforestation. Different people and groups continue to criticize the tardy implementation of these measures, but the real divergences between various sections have emerged over proposals aimed at controlling and removing encroachments in and around the lake. To counter the "shrinking" of the lake, LAWDA's plans include the uprooting of tree plantations, removal of large areas of floating islands, a moratorium on any new construction in and around the lake, the demolition of houses and eviction of many of the lake's inhabitants, and the relocation of houseboats to a fixed area in the lake. Concomitantly, the state government has developed plans for rehabilitating the lake-dwelling communities in other areas of the city. In 2007, it transferred 376 hectares of land to LAWDA for rehabilitation of the displaced families.

The officials of LAWDA have carried out some demolition drives within and outside the lake. A few families have been rehabilitated in a new residential project that has been heavily criticized for its poor quality. Nearly 400,000 trees and saplings were felled within the lake, but the stems and roots were not fully removed and many of them have regrown. The High Court has stayed the relocation of houseboats until the sewage treatment plants and other facilities are ready at the proposed sites. There has been scant progress in the removal of weeds, declogging of channels, and sewage disposal and treatment. LAWDA is widely accused of ecological mismanagement and corruption, even of willfully allowing new encroachments under political pressure.

The issue is deeper, however, as there has been little engagement or public consultation on the objectives of the clean up. Consequently, no collective vision has emerged for sustaining the lake's unique ecosystem and also, critically, its social world. The din over the "disappearing" lake has led to an aggressive cleanup monitored by the courts and a vociferous media, but critical questions have barely been addressed. What really should the cleanup, the "restoration" achieve? What should the lake be in the next ten, twenty, or fifty years? What is the "original" whose restoration is now sought?

The narrative of the "shrinking" lake is itself a contested one. Regular evocations of the lake as "disappearing" and "nearing extinction" advance the belief that there has been a progressive reduction in its expanse, but clear documentation supporting this assertion does not seem to exist. There is no one definite figure even for the lake's present expanse, and the figures cited from earlier periods to highlight the diminution of the lake are even more divergent. The varied estimates for the current size of the lake, proposed by different researchers, technical institutes, and government agencies, range from 11.5 to 25 square kilometers.[23] Concomitantly, there are no definitive numbers for the percentage of the lake that is open water and the percentage covered by gardens and habitation. LAWDA, citing a few of the studies, has asserted that the current expanse of the lake is 24.6 square kilometers. Some posit that the Dal's size in the twelfth century, when Kalhana's *Rajatarangini* was composed, was as large as 75 square kilometers, although 50 square kilometers is the more widely circulated figure. If these figures are to be believed, the lake has shrunk dramatically over the centuries and is now a fraction of its "original" size. The blame for this supposed shrinkage in expanse is largely attributed to "encroachment," mainly implying construction on reclaimed lake waters. The effect of these unsubstantiated figures has been so alarmist, however, that LAWDA has been forced to issue a public notice stating that there is no documented basis for the figures being circulated.[24] Awaiting any other "authentic" record that might emerge, LAWDA has stated that the figures from Walter Lawrence's Settlement Reports, from the end of the nineteenth century, should be accepted as the benchmark. Lawrence had calculated the total area of the lake as 25.86 square kilometers—with 18.21 square kilometers of that being "water surface" and 7.65 square kilometers as "land mass."[25] Thus, according to LAWDA, the overall expanse of the lake has decreased only marginally over the past century, from 25.86 square kilometers to 24.6 square kilometers.[26]

This assertion is contrary to what is visible, insofar as roads and houses have appeared on reclaimed areas around the lake. The larger consensus thus remains of a substantial shift in land use and land cover in and around Dal Lake over the past few decades. A study of lake size from 1980 to 2010—using remote sensing and geographic information system techniques—concluded that there is "a pattern of land transformation, where lake water is converted to marshy lands, which are subsequently converted for agriculture, orchards and residential uses."[27] A lot of this activity has happened around the margins of the lake and various sections of people are involved, but the discourse around the dwindling and "dying" lake apportions a large share of the blame to the lake dwellers, the Hanjis.[28] They are seen as "encroachers" and a strong focus of the "cleanup" is on removing them, their houses, and gardens from the lake, along with realigning their houseboats and restricting them to a small area. Blaming the inhabitants thus also helps absolve the state government and the municipal bodies, which have been unable to control the draining of sewage into the lake from most of the city.[29] More critically though, this discourse upholds the past and future vision of Dal Lake as distinct from its present—as an unobstructed stretch of water, not the amphibious landscape of today.

The Hanjis in turn have been fighting for their vision of the lake and their rights to stay on it. With their livelihoods threatened, they point instead to the many commercial and residential buildings around the lake, and the corruption and mismanagement of LAWDA.[30] As inhabitants of the lake for centuries, they say that they have learned to live sustainably within its ecosystem. They help keep the lake clean through the regular removal of weeds and silt from the water channels, by dredging out mud and spreading it on their fields as manure. The Hanjis contest the idea of the lake as a clear water body; in their imaginings, the Dal has "for most of its ecological history, been a combination of water and landmass."[31] With competing imaginings of the lake as a "heritage site" and a "natural spectacle," the Hanjis' contribution in "creating the spectacle" stands to be erased.[32] It is interesting though that the Hanjis themselves advance the heritage trope—for their lifestyles and their houseboats to assert their right to stay.[33] There are echoes here of the long debate around the conservation of protected areas, over the inclusion or exclusion of those who live within them.[34] Communities seek to legitimize their right to a place, to the ecology and landscape, by locating their origins and practices within it. However, counterarguments are presented and they are often blamed for the degradation of the landscape.[35]

For Dal Lake, the emphasis on the visual, on the gaze of the outsider, is so dominant that LAWDA repeatedly keeps water levels high to hide weeds and garbage. This is often done at the onset of the summer tourist season and for important political events. A notable instance took place in April 2005 during the inauguration of a bus service between Srinagar and Muzaffarabad, in Pakistan-administered Kashmir, which was hailed as an important "confidence building" measure toward resolution of the conflict.[36] Raising the water level causes many problems for the lake's residents, however, as vegetable gardens, platforms and anchors of houseboats, and shikara banks get submerged. It also prevents the required flushing of the lake's waters, as the outflow channels are closed to augment the water level.

The attention of the government and its agencies, the courts, the media, and many citizens remains directed toward a quick cleanup, but what does a cleanup imply for a semiaquatic landscape like the Dal? It must be emphasized that the stress on the lake's appearance as a gleaming body of water implies almost a metamorphosis: from an amphibious landscape to a clear body of water bereft of human inhabitants and their activities. However, this has never been clearly articulated or publicly discussed.

Tourism is seen as the backbone of Kashmir's economy, and its promotion is thus deemed a requisite for the well-being of the people. The often-touted figures state that about 50 percent of the valley's population is engaged directly or indirectly in the sector, and tourism's contribution to the state's gross domestic product is 12–15 percent. Some allege that these numbers are exaggerated and that foregrounding the role of tourism feeds into the state's narrative of the imperative for peace. For other Kashmiris though, it remains essential for more tourists to visit the valley and they suggest that the management of the Dal Lake should prioritize this end. A recent study calculated the gross annual revenue attributable to the Dal ecosystem to be about $62 million. However, it stressed, if the Hanjis are evicted and their economic activities curtailed, the financial loss would be offset by increased earnings from tourists, as more would travel to a cleaner lake and stay for longer periods.[37] Other lakes of Kashmir are also being "developed" for visitors, as indicated by an increased number of licenses issued for angling by the State Department of Fisheries. The department's focus seems to be on generating more revenue through tourism rather than on assisting the "conservation and development of lake fisheries which would enable the fishers to earn increased income."[38]

Concern that the rights of the dwellers not impede those who visit the Dal for recreation has long informed policies and plans for its conservation and development. A. F. Robertson, an anthropologist who worked on a consultancy in 1985, observed that his "commission reflected official anxiety that the people who lived and worked on the lake might thwart plans which were already in hand for its development as the region's principal tourist resort."[39] This anxiety has only deepened during the years of unrest, as tourism, long instated as the raison d'être of Dal Lake, has become intertwined with the politics of the conflict and its containment.

These complex issues with regard to "cleaning" the lake have become confused and there has hardly been any attendant discussion of the political situation, which permeates every aspect of life in Kashmir. The alarmist narrative has allowed the problems concerning the lake to be treated largely as "environmental" issues, divorced from the vortex of local and regional politics. Ecology nevertheless remains intertwined with politics, and particularly so in Kashmir where the conflict has deeply affected the governance and use of natural resources—especially of water.

DOMINIONS OVER NATURAL RESOURCES IN KASHMIR

In one of the many ironies of the conflict, the Kashmiris while demanding sovereignty have actually lost more autonomy over the years of the movement. In many aspects, the state government and its departments in Kashmir are less autonomous than other state governments in India. This is clearly evident in the management of its natural resources, of its forests, and of its networks of water.

Kashmir has extensive forests containing valuable timber. With collusion from various sections of the government, the illegal timber trade thrived during the violent decade, and widespread deforestation was reported throughout the valley. The catchment area of Dal Lake suffered too, although about half of its watershed forms the Dachigam National Park and is thus more protected. Still the presence of the Indian army in the park has had a major effect. Set in the Zabarwan Range, Dachigam is barely twenty kilometers away from Dal Lake. The altitude in the park varies from 5,500 feet to 14,000 feet, with coniferous broadleaf forests at the lower level, and sloping grasslands, meadows, and rocky outcrops as one moves higher. The high-altitude Marsar Lake is the source of the river Dagwan that gushes through the park and drains into the Dal. The forests of Dachigam were initially protected when the Maharaja of Jammu and Kashmir, Pratap Singh, declared it as a hunting reserve in 1910. This protection was also aimed at ensuring the drinking water supply of Srinagar. Dachigam was designated a National Park in 1981, land thus recognized as an area "of ecological, faunal, floral, geo-morphological or zoological importance" for the nation.[40] The park has a good population of black bears and leopards, apart from other animals and birds. However, it is most famous as the last refuge of the hangul, the Kashmir stag. The hangul is a subspecies of red deer, the only Asiatic survivor of the genus. Estimates suggest that only around 150 deer remain in the park. As the state animal, the hangul is also a symbol of Kashmir.[41] Apart from the environmental degradation that is affecting its habitat, the hangul is most threatened by the movement of people and domestic animals in the park. Villagers from adjacent areas get their animals to graze illegally in the rich pastures. A government sheep-breeding farm and a trout hatchery located within its boundaries lead to additional disturbance as the employees move through the park. The maximum interference is from the Indian armed forces, a battalion of the paramilitary Central Reserve Police Force (CRPF) that has been stationed in Dachigam since the beginning of the insurgency. The park is seen as a probable space for militants to hide and has purportedly been the site of many "encounters" with them, thus necessitating the army's presence and control. The army personnel cause a lot of disturbance with their movements and the barricades and barracks they have erected. Even more critical is the issue of the State Forest Department, which has the direct and primary responsibility for conservation and management of the park, having to cede control to the army. The Forest Rangers themselves now have to negotiate with the CRPF for access to parts of the park. Even ministers and senior government officials are formally required to obtain the Forest Department's permission for entry of their vehicles, but the vehicles of the armed forces are independent of the state's bureaucracy.[42] In May 2016, there were reports that heavily armed troops had conducted a night raid in Dachigam in complete violation of the regulations for National Parks. The office of the wildlife warden, having received no prior information, complained that the "sanctity" of the park was not respected.[43] It was an operation aimed at capturing the popular militant commander Burhan Wani, whose murder two months later led to massive unrest in the valley.[44]

Burhan Wani was gunned down in an "encounter" on July 8, 2016, and protests immediately erupted across the valley. In a saga all too familiar for Kashmir, the security forces responded with excessive force, fueling further anger and more protests. There had been weeks of protests in the valley in the summers of 2008 and 2010 too, but they were never so sustained. The Indian government and some observers blamed Pakistan for bolstering the unrest. To show its might the Indian army raided some posts across the border, but some sections of the Indian public and polity continued to demand further action against Pakistan. In response, the government began reviewing the Indus Water Treaty between the two countries, thus continuing the practice of deploying water as a strategic resource.

Part of the messy aftermath of the Partition was the sharing of water resources between India and Pakistan. Kashmir's geographical position as the source and catchment zone of some of the major rivers increased India's and Pakistan's stakes in it. After years of negotiation, the Indus Water Treaty was signed by the two countries in 1960, with the state government of Kashmir barely consulted. The treaty allocated the use of the water of the six major rivers of the Indus River system between the two countries. India was granted exclusive rights over the three eastern rivers (Sutlej, Beas, and Ravi), and Pakistan was granted exclusive rights over the three western rivers (Indus, Jhelum, and Chenab). They were allowed under certain, narrowly defined circumstances, to use the other's rivers. However, both countries have continued to argue that the treaty was more favorable to the other, and with many hydropower projects coming up to meet the growing demand for power, disputes have been taken to the International Court of Arbitration.

The Indus River and its tributaries flow through Kashmir before entering the complex canal system that irrigates most of the arable land of West Pakistan. This dependence of Pakistan on the waters of the Indus system also feeds internally into the militant discourse on Kashmir, emphasizing the need for Pakistan to maintain its claim on the valley.[45] Within India, the water from Kashmir's rivers is increasingly being harnessed to generate power.[46] The hydropower projects are controlled by the National Hydroelectric Power Corporation (NHPC), which shares a mere 12 percent of the energy with Kashmir as royalty. Successive state governments of Kashmir have sought to increase the proportion of NHPC royalties, but with little success.[47] Kashmiris often articulate their anger about India's denying them control over their own rivers. The issue of jurisdiction over hydropower projects is a point of conflict for the center with other states too, but embroiled in the Kashmir conflict, it emerges more potently.

Water is integral to Kashmir's ecology and economy, and the valley can be seen essentially as the basin of the river Jhelum, with its tributaries, streams, and canals. There are also more than a thousand small and large water bodies in the valley, fed by glacial melt and springs. Many of these are interconnected and the hydrological system supports a rich biodiversity. Apart from being sources of water for drinking and irrigation, the water bodies are critical for the livelihood of many—agriculturists, fishermen, and those involved in tourism. They are a source of life for Kashmiris, as exemplified by Dal Lake.

During the four months of unrest in the valley in 2016, the Dal Lake was a relatively untrammeled space where curfew could not be imposed and some semblance of daily life was extant. With the longest curfew in the valley's tumultuous history and the additional shutdowns imposed by separatists, there was barely any movement of vehicles and goods throughout the summer. Coincidentally, a bumper harvest of vegetables in the Dal helped sustain Srinagar's besieged residents.[48] The lack of cleaning during these months led to weeds and red algal bloom, which engulfed portions of the lake.[49] The political conflict thus continues to affect the landscape of the Dal in various ways.

Kashmir's ecology, economy, and polity are inextricably intertwined. The Dal Lake is a landscape, not just a physical environment; a landscape that reflects people's identities, aspirations, struggles, and power plays.[50] These imbue contesting narratives and historical imaginaries over the representation and use of its waters. Varying discourses in regard to "encroachment," "conservation," "cleaning" and "restoration" have emerged. None of these are unitary discourses though, as they draw from varied arenas. Advancing a singular alarmist narrative suggests disengagement from these multiple and often conflicting discourses. The effects of this disengagement are evident in the lack of influence of the various measures undertaken. The lake's condition has barely improved over the past fifteen years and the government; the courts and the media continue to bemoan its condition, pressing for yet more urgent action.

Some of the exigencies that have been discussed do qualify as alarmist, more so in a Western context: endemic pollution, fears of shrinkage, and a militarized territory. But does it make sense to view problems and conflicts in these terms? An alarmist outcry followed by a determined response falls short of the realities in Kashmir, and only feeds into the illusion that quick solutions are possible. The core question of how to nurture the unique landscape of Dal Lake remains unaddressed. The residents of Srinagar are not looking for a grand solution; the need is rather to accommodate daily concerns. Today despite the varied difficulties, life carries on much as usual in and around the lake. The Dal both as a place to live and as a space of leisure is a source of livelihood, indeed of life, to many; and while critical issues need to be addressed, treating it as a disaster area fails to address the most fundamental of these.

8

THE ADIVASI VERSUS COCA-COLA

A LOCAL ENVIRONMENTAL CONFLICT AND ITS GLOBAL RESONANCE

BERND-STEFAN GREWE

SHORTLY BEFORE THE END of 2005 an atmosphere of deep anxiety gripped Coca-Cola Company headquarters in Atlanta, Georgia. American students had started a campaign against the company, and in addition to the University of Michigan and New York University, another fifty colleges had terminated their contracts with Coca-Cola and banned the sale of Coke on their campuses.[1] The student protests had been sparked by reports of massive environmental damage caused by a large bottling plant owned by the company in the southern Indian village of Plachimada.[2] The loss of a few million dollars in sales as a result of the boycott was not particularly significant for the soda manufacturer, which had a total turnover of more than $23 billion.[3] Nevertheless, the corporation had been hit in a particularly sensitive area, since its more than one hundred years of success was due not only to the taste of its soda but, above all, to its brand image. Today, the Coca-Cola logo is still among the most famous trademarks in the world.

Most environmental conflicts of this sort were addressed, debated, and resolved at a local level. However, in this case the protest of a group of Indian village women attracted global attention. They achieved undreamed-of success when the bottling plant in Plachimada was closed, temporarily at first, and then for good. Their success was all the more astonishing since this was the case of a marginalized minority—women from the Adivasi community—winning out over an incredibly powerful global player. Adivasi are members of various indigenous ethnic minorities known in India as "scheduled tribes," and like the casteless group known as Dalits, they are at the bottom of the social hierarchy. Unlike large foreign investors, they do not have much lobbying power in Indian politics.

How can we explain their remarkable victory? How was this environmental conflict, which initially played out on a local level, able to morph into a global campaign against the Coca-Cola Company? What influence did the campaign have on the situation on the ground? And what does this say about apocalyptic tropes? This chapter seeks to explore these questions and to examine the influence of local actors on events that led to the closure of the factory.

THE EXPANSION OF THE COCA-COLA COMPANY IN INDIA

It was only in the 1990s that the Coca-Cola Company began to reestablish a foothold in India, after having pulled out of the South Asian market in 1977. After Indira Gandhi was voted out, the new Indian government pushed through a law (the Foreign Exchange Regulation Act) against the high-profile corporation, which was supposed to give the government more control over foreign companies, and stipulated that 40 percent of equity had to be Indian owned. Faced with this prospect, Coca-Cola decided to withdraw from the Indian market.[4]

It was not until 1992, after these restrictions were eased by the new, business-friendly Singh government, that Coca-Cola and its long-standing rival PepsiCo returned to India. As a result of shrewd takeovers and the construction of manufacturing plants, the two soda giants controlled almost 80 percent of the market for soft drinks by the turn of the millennium, their main consumers coming from the rapidly expanding urban middle class.

The conditions for large-scale investment were excellent because since 1991 India's central government in New Delhi and the governments of several Indian states had been competing to attract foreign investors. The costs of producing drinks were also lower than elsewhere: a water law dating back to the British colonial period allowed landowners to use as much ground water as they wanted from wells that were on company-owned land. Unlimited usage was allowed without any additional costs, and the price of labor was also very low. In the first ten years after operations were restarted in India, the market for soft drinks grew by an average of 25 percent a year, producing exceptional returns that could be transferred back to Atlanta.[5] As a result, both drink corporations expanded their production and in 1998 Coca-Cola put in an application to open a manufacturing facility in Plachimada, in the southern Indian state of Kerala. The Panchayat (village council) of Perumatty approved the application, which promised to create new jobs, and 141 permanent positions were opened up, in addition to a further 220 fixed-term roles.[6]

Plachimada, Kerala, had been carefully selected on account of its location. The inland region in the south of the subcontinent often suffered from droughts when monsoon rains failed. However, the district of Palakkad, where the village was situated, was fertile and had plenty of water; it was even seen as the "breadbasket" of the state. The Coca-Cola Company had taken a systematic approach to identifying a suitable location for the bottling plant and the wells that it needed. The firm had satellite images scientifically analyzed and discovered a large aquifer, which it hoped meant that the plant could use large volumes of water on a permanent basis.[7]

The bottling plant in Plachimada was constructed and from March 2000 onward it bottled up and transported away around 500,000 liters of soda every day. The water for the plant was extracted from about 60 deep wells; 2.7 to 3.8 liters of water were required to make each liter of soda. Information regarding actual usage is conflicting and varies between 500,000 liters of groundwater (according to Coca-Cola) and 1.5 million liters (according to environmental organizations).[8]

The first residents' complaints about the company's use of drinking water arose after just a few months. These were followed by protests, which spiraled into Coca-Cola's greatest ever crisis in India. Demonstrators and the international press blamed the soda manufacturer for four environmental problems that arose between 2002 and 2005. First, the bottling plant in Plachimada was said to have polluted residents' drinking water with hazardous runoffs, rendering it unusable. Second, excessive extraction of water to produce soda and wash bottles was said to have caused a massive drop in the water table, leading many wells to dry up. Third, Coca-Cola was accused of having given sludge contaminated with poisonous cadmium and heavy metals to farmers to use as fertilizer. Fourth, Indian cola itself was said to be a health hazard because it contained a high proportion of pesticides. Three of these issues were centered on Plachimada; the fourth simply exacerbated the existing crisis.

WATER POLLUTION AND OVERUSE: ENVIRONMENTAL PROBLEMS IN PLACHIMADA

The protests in Plachimada began after problems caused by the plant began to mount up. At first, residents in the area observed that their wells had become briny and had turned a milky-white color. They noticed a smell that resembled a mixture of stale palm wine and kerosene. Locals complained of diarrhea, unusual stomach pains, rashes, and increased hair loss: "Our water was pure before the factory opened. We never had any problems. Now we can't even bathe in the water. Our hair clumps together and falls out, even on [the] head of my baby. We can't drink this water. It hurts (pointing to her stomach). The workers in the fields come home and find their feed [sic] and legs covered in rashes. They ruined our water and land, and they don't care."[9] On the grounds of taste alone, the well water was no longer suitable for drinking and cooking. Women now had to walk several kilometers a few times a day to fetch the drinking water they needed. This also meant a loss of income for rural women and their families, because carrying water over these distances made it impossible for them to take on paid work.[10] The residents believed the pollution was caused by wastewater that had not been treated to remove residual dirt and chemicals and was being pumped into dried-out wells. This sparked local protests and was a catalyst in their development.

The bottling plant was connected to a second environmental problem that was even more serious for residents across the entire region: the lowering of the water table. Locals blamed the factory's heavy extraction of water for causing the water table to drop from 45 meters to more than 150 meters. Since farmers needed these wells to water their fields, and were thus dependent on them for their survival, this was a real crisis for them. Some wells even dried up altogether—there were reports in the regional press of 250 dry wells. The lack of water in an area where it had always been plentiful meant that harvests rapidly declined. Within the next three years, the yield per land plot sank from an average of 50 sacks of rice to just 5, and the yield of coconuts dropped from 1,500 to 200.[11]

For local farmers and landowners this meant a dramatic loss of income. But the effect was even more severe for the majority of the population, who owned little or no land and who could no longer find work as farm laborers, meaning that they made no money at all. These farm laborers were predominantly casteless people (Dalits) and Adivasi, indigenous groups of varying ethnic origin sometimes referred to as "aboriginals" or "scheduled tribes."[12]

Initially, the residents of the nearest village commissioned a private, independent laboratory to confirm that the well water was unfit for consumption. A state-accredited laboratory attested to this fact and suggested in its report that contamination could have been caused by the neighboring bottling plant.[13] Next, the victims collected signatures and sent petitions to the local district collector, the water authority, the Pollution Control Board, the Panchayat in its role as local authority, and to elected members of parliament. The residents did not receive a single response to these petitions. They then founded the "Coca-Cola Virudha Janakeeya Samara Samithy" (Anti–Coca-Cola People's Struggle Committee). In the tradition of Mahatma Gandhi, they focused on illegal but nonviolent action.

Organized resistance began in Plachimada on April 22, 2002. It was inspired by the example of C. K. Janu, an experienced politician and an Adivasi woman, who had fought for years with some success for land to be returned to and distributed among the Adivasi.[14] The resistance movement was primarily driven by women. The Adivasi invited Janu to the founding of the anti–Coca-Cola movement. On International Earth Day around five hundred women and a few men gathered for a sit-in in front of the gates of the bottling plant, thus blocking lorries from entering. The next day police units violently broke up the demonstration and arrested some of the activists.[15] Despite intimidation and additional arrests, the women continued their daily protests over the following months. Official reactions still failed to materialize, but the women maintained their daily vigil opposite the factory gates and demanded that the plant be closed and the victims compensated.[16]

The factory's management categorically denied that there was any pollution and immediately set up a committee to protect the five hundred jobs that were supposedly at stake. At the same time, they threatened to dismiss eighteen temporary workers from Plachimada, who had taken part in the protests as local residents, if they did not sign a declaration asserting that Coca-Cola was not responsible for the water pollution.[17] Coca-Cola often implied that protests of this sort were "politically motivated."[18] In fact, the concerns of the residents were ignored by politicians of all parties as well as by the state authorities.[19]

The protest movement, which was initially supported only by Adivasi organizations, tried to attract additional attention by organizing a protest march to the Panchayat, or village council. This form of local government, which almost exclusively represents wealthy farmers and businesspeople, is viewed in India as notoriously corrupt. Up to this point, the Panchayat of Perumatty had not paid any attention to the protests. Now the women symbolically cleaned the council building with brooms, and at the following rally the council's silence was severely criticized. This step made an important contribution to the campaign, in that the Panchayat took on the issue and made its own assessment of the situation.[20]

Political parties did not become involved in the conflict until November 2002, when the Communist Party of India, or CPI (Marxist), was the first to declare its solidarity with the protestors. When coverage of the issue in the regional media increased, the opposition party seized the opportunity to use the protest movement as a platform for its political struggle against the ruling center-right coalition in Kerala, and to use the media spotlight to agitate against U.S. imperialism and global capitalism. Other leftist parties likewise began to declare their solidarity. Environmental organizations, the fishing union of Kerala, and other opponents of globalization also promised to lend their support and organized a solidarity committee and a number of protest marches in Kerala. These marches were targeted firmly against economic globalization, of which Plachimada was invoked as a symbol.

A Kerala parliamentary commission visited the factory two months later, debated the environmental issues with the management, and ordered a scientific evaluation to be conducted to investigate the accusation that groundwater was being overused. The parliamentary representatives did not speak to the protesters at the factory gates.

Attitudes toward Coca-Cola in India worsened considerably when the United States invaded Iraq in 2003 in order to find supposed "weapons of mass destruction." The war, which Muslim Indians were not alone in viewing as illegal and unjust, led to calls to boycott American products, in particular Coca-Cola, which was identified with the American way of life more than any other brand. This identification with the United States, which had previously benefited sales, now had a negative effect on them and strengthened people's dislike of the company. Partially violent protests ensued; some factory facilities belonging to PepsiCo and Coca-Cola were attacked and cola bottles were publicly smashed in the street. In Plachimada, however, the same outrage did not boil over; the protest continued peacefully.

TOXIC SLUDGE AS MANURE, AND PESTICIDE IN THE COLA

Coca-Cola's predicament became considerably worse in the summer of 2003. There was a groundswell of support for the protesters from across the country after a report by the BBC about the conflict in Plachimada was broadcast at the end of July 2003 on Radio 4's *Face to Face* program. The reporter John Waite traveled to Plachimada and learned that Coca-Cola was distributing dried sludge to local farmers to use as free fertilizer. The sludge was often deposited outside the plant next to irrigation channels, and rain washed it into rice fields, water channels, and wells. The BBC reporter collected samples from wells and from the dried sludge and had them analyzed by the University of Exeter. The head of the laboratory gave his conclusion in no uncertain terms: "What is particularly disturbing is that the contamination has spread to the water supply—with levels of lead in a near-by well at levels well above those set by the World Health Organization."[21] The vice president of Coca-Cola India, Sunil Gupta, who was also interviewed, vigorously disputed this result. He referred to contrary scientific findings, but he was unable to present any of these to the BBC.[22]

Thus, Coca-Cola was blamed for a third environmental problem in Plachimada. Coca-Cola indirectly admitted to this form of pollution when it referenced the "facts" in a public relations (PR) campaign: "The 'sludge' or bio-solid is the end result of the waste water and water treatment processes and is made up of organic and inorganic material. The use of bio-solids as a soil amendment is not an uncommon practice around the world and within the Coca-Cola system, including in the US . . . the Kerala State Pollution Control Board, in a detailed study, concluded that the concentration[s] of cadmium and other heavy metals in the bio-solids are below prescribed limits and, therefore, are not considered hazardous."[23] Thus, Coca-Cola professed to having given production waste to farmers and represented this as common practice. According to the protesters and to the scientific findings, this was by no means just a case of downplaying the use of organic sludge. Rather, Coca-Cola was trying to suggest that it had stayed under the required limits and therefore the practice should not be considered dangerous. This was simply a lie: the proportion of hazardous substances was many times higher than even the low Indian safety standards permitted. When it comes to matters of environment and health protection, Indian law is still underdeveloped. Environmental organizations have been calling for years for the law to be updated, but the business-friendly government, with an eye on foreign investors and domestic producers, has refused to grant this request. Thus, India allows drinking water to be polluted to a level that is many times higher than what is permitted in the United States or the European Union (EU).

Just two weeks after the radio program was aired, the Kerala State Pollution Control Board published a report that found even higher levels of cadmium and lead in the sludge than the English laboratory had detected; the proportion of cadmium was over 200 milligrams per kilogram (mg per kg).[24] By way of comparison, in the EU the maximum amount of cadmium that can be present in agricultural biosludge is 1.5 micrograms (µg) per kg, and the maximum amount of lead is 100 µg per kg. The units here are crucial: the Indian unit of measurement of 200 mg per kg is equivalent to 200,000 µg per kg. In other words, the amount is more than 100,000 times above the EU limit. Because cadmium is highly toxic, the amount allowed in drinking water in the EU was lowered in 2011 from 5 µg to 3 µg per liter (in India it is 0.01 mg, equivalent to 10 µg); 10 µg of lead per liter is allowed (in India it is 50 µg per liter).[25]

Suddenly, just one month later, a senior representative named Indulal, a member of the same environmental board (Kerala State Pollution Control Board; KSPCB), made the surprising declaration that these results did not represent a health hazard and that Indian standards had not been broken. As can be seen in the quotation above, Coca-Cola later used this obviously false statement without referring to the published data, which showed that the levels of toxins were clearly too high. Indulal's extremely one-sided views, and the disregard for all facts that were potentially damaging to Coca-Cola, suggested a case of corruption. These suspicions led the state Anti-Corruption Bureau to investigate, and Indulal's four houses were searched. He subsequently resigned from his post with the KSPCB and began working as a consultant for Asia Bank.[26] More than likely, this was a case of bribery; his interpretations differed far too obviously from all the findings of his own agency and from those of other laboratories. Opponents of Plachimada reported other bribery attempts by the Coca-Cola Company. A spokeswoman for the demonstrators testified in a documentary film that the Adivasi were not only offered houses, electricity, and schools if they ended their protest, they were even offered education for their children. According to statements by Krishnan, the former spokesperson for the Panchayat, the council was offered 380 million rupees (around €6.8 million at the exchange rate at the time) for "community engagement" if an agreement could be reached on a new operating license.[27] This collection of unusually generous proposals immediately suggests that the company was trying to remove an obstacle by indirect means. But if offers really were put forward in this form, the question still arises as to why the mostly poor Adivasi turned down these opportunities. At the same time, it should not be overlooked that corrupting supposed experts and elected committees was clearly part of Coca-Cola India's repertoire of business strategies.[28]

As it was, the KSPCB banned Coca-Cola from distributing sludge and it ordered that the sludge that had already been handed out should be recollected and safely stored on company property.[29] Nevertheless, this did not prevent Coca-Cola from continuing to assert that the company used the latest wastewater technology and complied with all regulations: "The technology our wastewater treatment plant uses is among the most advanced in the world. The technologies are also equivalent to most Coca-Cola bottling plants in the United States and Europe. Further, our effluents comply with standards and norms set by the Kerala State Pollution Control Board. We constantly monitor the quality of the effluents to prevent pollution."[30] For the affected residents in Plachimada and for the solidarity movement, such statements must have been a slap in the face. They considerably strengthened the sense of outrage about the company's behavior. India's national press now intensified its reporting on the conflict in Kerala.

In the midst of the controversy over production waste and wastewater, a report by the Centre for Science and the Environment (CSE) in New Delhi (August 5, 2003) unleashed a fresh torrent of negative press reports about Coca-Cola in India. This nongovernmental organization (NGO) received financial support from the Heinrich Böll Foundation, MISEREOR, and from the European Commission. It had sampled twelve different soft drinks from three different regions and tested their pesticide content (Lindane, DDT, Chlorpyrifos, Malathion). The results were alarming. Pepsi-Cola exhibited concentrations of pesticide that were around thirty-seven times higher than European safety standards, which was used as a benchmark. Coca-Cola, which was likewise tested, showed concentrations about forty-five times above the limit. By comparison, the samples of both types of cola taken from the United States showed no pesticide residues at all.[31]

The effect was immediate. The Indian national parliament banned the sale of cola from its cafeterias. Several schools and universities followed suit and many among the Indian middle class were deeply worried. Sales of cola abruptly dropped by a third, and in the days after the CSE report was released the American share price of Coca-Cola dropped from $55 to $50.

For decades, the two soft-drink giants Coca-Cola and PepsiCo had competed bitterly in markets around the world. Now they appeared in a joint press conference in which they criticized the CSE's methodology and offered up their own, differing test results. In the ensuing period they placed full-page advertisements in leading Indian dailies to present their own studies.[32]

When faced with corporate scandals, it is common practice for companies to cast doubt on studies that have produced undesirable results. After being criticized by Coca-Cola and PepsiCo, the CSE reexamined the information that had been published by both manufacturers and stated that in the case of Coca-Cola the data did not actually relate to the soda, but to mineral water that was subject to much stricter controls (Kinley mineral water), while PepsiCo made the mistake of saying that the water they used had fewer pesticides in it than the soda that was produced in their factory. Moreover, the data came from other bottling plants, and the companies themselves had done the tests.[33] The PR campaign did not produce the desired result: one state even considered introducing a ban on the soda.

The misinformation in Coca-Cola's PR material and the conflict in Plachimada, which attracted additional attention as a result, proved to be a danger to the company's carefully constructed brand image. The corporation showed itself to be more concerned about this image than it was about the health of the residents of Plachimada or the Indian consumers of its products. A later test conducted in 2006 proved that levels of pesticides in the soft drinks were still very high.[34] Curiously, the English Guardian newspaper reported (November 1, 2004) that some Indian farmers now used the cola itself as a pesticide because it was said to be many times cheaper than similar pesticides sold by Monsanto, Shell, or Dow. However, an agricultural expert who was interviewed for the article did not connect this practice to the amount of pesticide in the soda. Instead, he pointed out that farmers had long used sugar water to attract red ants, which then fed off insect larvae.[35]

The crisis now continued to intensify for the cola manufacturers. The World Social Forum (WSF), held in Mumbai, January 16–21, 2004, gave representatives of the Plachimada Solidarity Committee the opportunity to meet with other NGOs and obtain their support for a World Water Conference. More than sixty activists who had been at the WSF met again in Plachimada, January 21–23. Among them were internationally renowned critics of globalization such as France's José Bové, the Canadian Maude Barlow (author of *The Blue Gold*), the activist Ward Morehouse (d. 2012), who wrote several books about the chemical catastrophe that took place in the Indian city of Bhopal in 1984 and organized an international campaign against Dow Chemicals. Also present was the Indian ecofeminist and recipient of the Alternative Nobel Prize, Vandana Shiva. It was at this headline-grabbing event that the "Plachimada Declaration" and the slogan "Quit India," which evoked memories of the Indian independence movement, were announced. "We, who are in the battlefield in full solidarity with the Adivasis who have put up resistance against the tortures of the horrid commercial forces in Plachimada, exhort the people all over the world to boycott the products of Coca Cola and Pepsi. Coca Cola—Pepsi Cola 'quit India.'"[36] This declaration treated the conflict over water as an issue of basic rights: "Water is the fundamental right of all people. It has to be conserved, protected and managed. It is our fundamental obligation to prevent water scarcity and pollution and to preserve it for generations. Water is not a commodity. We should resist all criminal attempts to marketise, privatise and corporatise water." Because the conference targeted not only Coca-Cola but also PepsiCo, the final rally took place in front of the PepsiCo factory in Pudusseri, which was only 40 kilometers away.[37] Media attention was guaranteed: left-wing and liberal news outlets in Europe (including The Guardian, Spiegel-online, Le Monde, Le Monde Diplomatique, and L'Express) were reporting more and more frequently on the scandal,[38] even though the conservative press such as the Frankfurter Allgemeine Zeitung, Die Welt, the Times, the Wall Street Journal, the New York Times, as well as the Süddeutsche Zeitung, Die Zeit, and El País, initially did not cover the issue.[39]

Because sales of Coca-Cola (and PepsiCo) in India were now plummeting, the company commissioned a PR agency to improve its brand image and boost its revenue in the country. In the ensuing period, neither PepsiCo nor Coca-Cola released any information about movements in their sales figures.[40]

The agency Perfect Relations advised Coca-Cola to form an independent, external advisory board.[41] When this board officially began work in December 2003, Coca-Cola also created an Indian Environment Council with the former chief justice of India, Bhupinder Nath Kirpal, as its spokesman. Kirpal attacked the rulings that the High Court of Kerala had made against Coca-Cola.[42] Environmental activists criticized the new environment council and environmental initiatives as nothing more than a PR exercise to distract people from the ecological problems that the company had created.[43]

THE FIGHT OVER THE CLOSURE OF THE PLACHIMADA PLANT

In parallel with this "war of the experts" and these public disputes, Coca-Cola was also in conflict with the authorities over whether the bottling plant in Plachimada would be allowed to continue operating.[44] At the end of March 2003, the company's temporary operating license expired and needed to be renewed. At the beginning of April 2003, almost a year after the daily protests had begun, the Panchayat announced that it did not intend to grant an extension to the plant's license. The Panchayat said that this was due to the rapid depletion of groundwater resources and it invited Coca-Cola to attend a hearing. However, Coca-Cola's management did not take this opportunity to settle the matter or to make any binding concessions. No representative of the company ever appeared before the council.[45] The Panchayat set Coca-Cola a deadline to submit the requested evidence about water extraction and waste disposal, otherwise the plant would have to cease operations from May 17.

Instead of responding to the accusations, Coca-Cola instructed its lawyers to file a case with the High Court in Kerala (the highest court in the state) to contest the denial of the permit. However, the court referred the issue back to the Local Self-Government Department. The administration then ruled that the Panchayat had overstepped its authority.[46]

Immediately afterward the state environment board, the KSPCB, also became involved in the dispute over the permit. It decreed on August 7, 2003, that because of the high levels of cadmium in the sludge, the factory was no longer allowed to distribute its waste. The Panchayat then used this decree to demand on September 9, 2003, that Coca-Cola provide evidence on the matter. However, instead of handing over this evidence, Coca-Cola once again filed an appeal with the High Court.

A few days later, the communal authority ordered the Panchayat to commission a team of experts to carry out a detailed investigation so that a decision could be made about the operating license. However, because this would be very expensive, the Panchayat now filed an appeal against this order with the High Court. On December 16, 2003, a High Court judge ruled that the Panchayat had the right to refuse to provide an operating license. Coca-Cola appealed, arguing that the decision to suspend the license was arbitrary and biased and that it did not rest on scientifically proven facts. However, the court judged that drinking water was a public right, and could not be overused by a private party. Coca-Cola was permitted to use the groundwater for one more month, but after that it was told it would have to source the necessary water by other means.

Coca-Cola also appealed against this decision and the plant continued to operate without a valid license. The High Court then appointed a new chamber with two judges who reversed the earlier decision, ordered the Panchayat to grant the operating license, and stop obstructing the plant's operations.[47] Coca-Cola subsequently purchased private groundwater and transported it to the plant by tanker truck. A segment of the movement became radicalized and turned significantly more militant; a group of young men and women hijacked a truck and emptied the water into nearby fields.

Intensive lobbying by the Plachimada Solidarity Committee led the government of Kerala to look into the water shortage. It was concluded that seven of the state's fourteen districts were suffering from drought. Among them was the district of Palakkad, which contained the municipality of Perumatty and the village of Plachimada, and previously had plenty of water. In February 2004, the government prohibited Coca-Cola from extracting groundwater until the middle of June, when the next monsoon rains were due. As had now become routine, Coca-Cola again filed an appeal but the High Court ruled that the government's decision was legitimate. Finally, on March 9, 2004, operations at the plant were halted, for the time being.

Meanwhile, however, the legal dispute over the operating license continued. In 2004 the Panchayat again turned down the five-year license, the government reversed the decision on the basis that it overstepped the Panchayat's authority, and the High Court ruled that, at least temporarily, the plant was not allowed to pump groundwater, but that it was also not permissible for its operations to be halted.

A few months later, in September 2005, the soft-drink manufacturer applied to the KSPCB environment board for a new environmental accreditation, in order to be able to continue its production from January 2005. This accreditation was denied, first, because the company had not provided any information about the origin of the cadmium found in its waste, and second, because it had not fulfilled the requirements of a commission of the Indian Supreme Court. The commission had, without success, instructed Coca-Cola to distribute drinking water to affected residents and to install a more effective wastewater treatment plant.

In the legal dispute over the license, the team of experts that had been called for by the High Court finally delivered their results in February 2005. On the basis of this report, the court judged that Coca-Cola should be allowed to continue using 500,000 liters of groundwater a day if enough rainwater had fallen that year; if there was less rain, the company would have to reduce the amount of water it extracted accordingly. Neither party was happy with this decision, and they contested the work of the commission—Coca-Cola because it was not allowed to pump as much water as it wanted on a continuous basis, and the protest movement because the decision had not taken into account the question of water pollution.[48] The court reversed the ban on operations in Plachimada, and once again ordered the Panchayat to grant an operating license to Coca-Cola within a week, if the company could present the required evidence as well as the approval of the Pollution Control Board (KSPCB).

In the next round of the legal battle, the Panchayat refused the operating license on the grounds that documentation was lacking (April 26, 2005), and Coca-Cola once again went to the High Court. The court ruled in June that the Panchayat must provide a temporary license for three months, within which time the company had to produce the licenses and permissions. As a result, the plant resumed operations on August 8, 2005, and again applied for clearance from the KSPCB. This environmental authorization had been denied the year before on the basis of unfulfilled requirements relating to the cadmium issue and to the commission of the Supreme Court, and the order was given to immediately cease operations at the plant.[49] We can only speculate about the extent to which public opinion ultimately affected the decisions of the courts and the authorities; there are indicators on this front, but there is no firm evidence.

This process, whereby Coca-Cola tried to use litigation to win a permit that had been denied to it, first by the Panchayat (whose ruling was successfully appealed against at the High Court) and then by the KSPCB, was repeated in the autumn of 2005 for a third time, with the same result. In January 2006

Coca-Cola began to publicly discuss the closure and relocation of the plant for the first time, and it made an application for a transfer to a different district.[50] The plant never resumed operations. Nevertheless, the conflict was by no means over.

THE EFFECTS AND POLITICAL CONSEQUENCES OF THE CONFLICT

Coca-Cola could not prevent or mitigate the crisis in India through using PR and the counter-opinions of experts, or through appealing to the courts, or through corruption. Instead, public pressure on the corporation increased even after the closure of the bottling plant. In 2006, following a change of government in Kerala (the Left Democratic Front, led by the Communist Party of India [Marxist],[51] took power), the sale of Coke and Pepsi was completely banned in the state. However, the ban on cola had to be reversed after a month because of procedural errors in the High Court's instructions. Unsurprisingly for a Marxist party, the CPI (Marxist) had long aligned itself with critics of globalization and taken up a dedicated stance against multinational corporations and economic globalization.[52] The party was the strongest force in the new, left-wing government, so Coca-Cola had little hope of negotiating successfully to restart the licensing process. The plant in Plachimada remained closed and was not to be reopened.

Although this put an end to the actual crisis in Plachimada, criticism of the soft-drink corporation persisted. Numerous universities in the United States terminated their contracts with Coca-Cola, students organized solidarity demonstrations and set up Web sites that provided information about the farmers' struggle against Coca-Cola. A local protest morphed into an international campaign. Water activists and environmental campaigners allied themselves with Adivasi organizations and with opponents of globalization from all over the world.

The international movement against Coca-Cola's business practices was no longer being triggered and directed by the traditional print media or television; instead, it was dependent on the Internet and on networks between different NGOs. Numerous linked Web sites shared the news of the Adivasi struggle against the powerful Coca-Cola Corporation. For journalists, this David versus Goliath narrative was an attractive story that gained even more attention as a result of the pesticide scandal. Activists managed to interest the press in this story again and again, and campaigners themselves were sometimes also very active in the media (for example, Bijoy, Mathew, and Shava).

Of course, Coca-Cola was also very energetic in its PR, and the company had access to a massive advertising budget that enabled it not only to place full-page ads full of counter-information in newspapers but also to create ads for TV and cinema. It cannot be denied that, with an eye to their lucrative advertising contracts with the soft-drink giants, some editorial departments rather shied away from publishing critical reports about Coca-Cola and PepsiCo, particularly in the early stages of the conflict. Widespread, critical reporting of the events only began after the population, and thus the media's readership, had been mobilized.

The political leanings of different media outlets were reflected in the reports they published and the information they selected. The *Wall Street Journal* or *USA Today*, for example, showed themselves to be very one-sided in their support of the listed company and were happy to publish Coca-Cola's press releases or relevant interviews, thus offering the corporation a platform for its PR. Criticism of Coca-Cola in India was not to be found here.[53]

The Plachimada crisis is an example of how the new structures of communication enabled by the Internet radically altered public perception. Public opinion could no longer be controlled as easily as it had been in previous decades. As Coca-Cola's crisis in India showed, the company's whole PR push was of little use. Thus, in an article in the *Wall Street Journal* the company lamented that its own Web site (www.cokefacts.org, which has since been taken down) only had around 800 visitors in the course of a month, whereas activist Amit Srivastava's Web site had over 20,000 hits in the same period.[54] His Webpage (www.indiaresource.org) reported regularly on the anti–Coca-Cola campaign in various Indian states and in the United States. Likewise, the Web site started by Ray Rogers (www.killercoke.org), which was originally intended to raise awareness about crimes committed against Colombian trade unionists, also contributed significantly to initiating and spreading student protests.

These Web sites also helped other communities who lived near Coca-Cola bottling plants and were affected by water shortages and pollution to inform themselves and to imitate certain tactics or deploy them even more effectively. The example of Plachimada spawned a number of imitators, and similar conflicts with Coca-Cola or PepsiCo followed in other locations. In Mehdiganj, near Varanasi (in the state of Uttar Pradesh), in Kaladera near Jaipur (in Rajasthan), in the district of Thana (in Maharashtra), and in Sivaganga and Gangaikondan (both in Tamil Nadu), similar protest movements sprang up against bottling plants and administrative resistance increased.[55]

To put a stop to the protests and the ban on Coke at the University of Michigan, which was damaging the company's reputation, the university administration and Coca-Cola agreed to have the Indian bottling plants assessed by an independent institute. The Energy and Resource Institute (TERI) was commissioned and it submitted its report in January 2008. TERI's results contradicted some of the protesters' arguments and found that the quality of the wastewater produced by the plant met the required criteria and was up to company standards. It judged the water as follows: "On an overall basis, the quality of the treated wastewater indicates effective treatment for most of the parameters vis-à-vis prescribed pollution control board discharge standards."[56] The qualifier that only most, but not all, parameters were up to standard was later overlooked in Coca-Cola's portrayal of the report. Representatives of the protest movement immediately pointed out that Coca-Cola was among the institute's most significant donors and financed several of its projects, and thus they called the objectivity of the study into question.[57] However, it was not the institute's intention to whitewash Coca-Cola's actions; on the contrary it confirmed the accusation that the company was overusing numerous groundwater deposits, and it criticized the low Indian water standards, which a company such as Coca-Cola ought to exceed. TERI advised that one of the six plants in the study (Kaladera) should be closed completely because the groundwater resources had already been exhausted.[58]

In the meantime, Coca-Cola had adopted a noticeable change of tone. Instead of characterizing the results of all other tests as false accusations, as it had in previous years, spokespeople for the firm now indicated that there was still much to do, and that the company intended to make greater efforts to improve its record on water usage. Therefore, the company said it would collect and use more rainwater.[59] The management of Coca-Cola India suggested a new direction in the way it approached water usage: "At The Coca-Cola Company, we are transforming the way we think and act about water stewardship. It is in the long-term interest of both our business and the communities where we operate to be good stewards of our most critical shared resource, water."[60] Coca-Cola wanted to use the example of the bottling plant

in the drought-prone village of Kaladera (Rajastan) to show that it was serious about saving water. The company stated that it had managed to reduce
its water usage there by half. (This initially impressive saving was, however,
principally achieved by replacing reusable glass bottles with disposable ones
made of polyethlylene terephthalate (PET) that did not have to be washed
out, and thus contributed to a growing mountain of plastic waste). In addition, Coca-Cola said that reservoirs, which it had financed, were collecting
rainwater, equal in volume to fifteen times the amount of water that the plant
was using, and diverting it back into the groundwater system. In Coca-Cola's
view, this made it possible for the factory in Kaladera to continue operating,
even though TERI had already advised that it should be closed. The University of Michigan remained skeptical and had Coca-Cola's statements tested at
the site. The analysis revealed that some of these water-harvesting facilities
were in a terrible state and unable to function. Contrary to what had been
agreed with the university, Coca-Cola had not installed a measurement system so that the amount of water collected could be monitored. At the same
time, the company's own calculations, which were supposed to prove that
water reserves were being replenished, were kept secret. The study concluded:
"The company is not holding its own explicit promises."[61]

THE POWER OF THE LOCAL? LOCAL PLAYERS
IN A GLOBAL CONTEXT

The Adivasis' local protest in Plachimada grew between 2002 and 2005 to
become an international campaign against the Coca-Cola Company that
drew the support of numerous NGOs. The question arises as to how much
influence the local actors actually had on the process that led to the final closure of the plant and how the "power of the local" can be seen in this context.
Ultimately, this particular conflict revolved around a specific dispute between
a local community and a global player, a corporation that operated in countries all over the world. Thus, it can be seen as a particularly appropriate case
through which to consider connections between the global and the local.[62]

The example of Plachimada reveals how the local issues of an ethnic
minority, the Adivasi, evolved into something completely new through the
involvement of an ever greater number of actors. For the Adivasi women
and for the farm laborers it was a question of survival; the dispute was over

their everyday drinking water and their own meager incomes as well as their constitutional rights. As antiglobalization and anticapitalist rhetoric became embedded in the discourse, the conflict became a fight between "David and Goliath," between underprivileged people in a developing country and an enormous corporation with a world famous brand. Plachimada, which was incorporated into an international discourse about globalization and its consequences, became a symbol for a successful struggle against economic globalization and a model for many victims in similar conflicts.

The ground for this global resonance for a local environmental conflict had been prepared for many years by an international debate on conservation and protection of drinking water supply but also on the privatization of water supply that was perceived as a danger by many environmentalists and NGOS. There was a growing international awareness that several hundred million people lacked access to clean drinking water, international think tanks such as the World Water Council (founded in 1996) promoted the idea of a sustainable use of water. In 2003, just a few months after the Plachimada protests had started, the United Nations Committee on Economic, Social and Cultural Rights declared: "The human right to water entitles everyone to sufficient, safe, acceptable, physically accessible and affordable water for personal and domestic uses."[63] In 2010 the Human Right to Water and Sanitation was recognized by the United Nations General Assembly. Although droughts in the state of California had not been rare, its population had to face two severe droughts from 2007 to 2009 and again from 2011 to 2017. Both reached California several years after the Plachimada struggle and thus did not influence public reaction in the United States to this case. Plachimada was not an apocalyptic trope, but a strong symbol of the necessity to think differently about the distribution and use of water in developing as well as developed countries.

As the protests spread and more and more people became involved, the Adivasi gained support in their struggle. At the same time, it looked as if they had lost the autonomy to interpret their own situation. The conflict was now reframed according to preexisting international discourses about the consequences of globalization. This meant that there was a focus on some political issues that the Adivasi were initially not concerned with at all. On this point, the question also arises as to whether, from a cultural sciences standpoint, we ought to speak of a "power of discourses" and in a social sciences context of a "power of social networks" and an "impotence of the local."

This point can explain the shifting areas of focus within the anti–Coca-Cola movement. The campaign, which fought at first against water pollution and water scarcity, became increasingly directed toward preventing the seizure of scarce resources by "monopoly capitalists" and the negative effects of economic globalization on impoverished rural communities. However, this shift does not mean that a preexisting globalization movement simply appropriated the protest in Plachimada and used the Adivasi. In India, critics of economic globalization could draw on a long tradition of anticolonial and anti-imperialist discourse. Such discourse is familiar to all Indians as one of the most important national narratives in the struggle for political and economic independence. As a form of knowledge that could be mobilized, it could easily be transferred from the British to the Americans.

Particularly in the case of the CPI (Marxist) there is a suggestion that the party was to some extent an outsider that adopted the conflict to achieve its own political goals and to please the electorate, at least on a symbolic level. When the communists took control of the government, the factory in Plachimada remained closed and sales of Coke were even temporarily forbidden. Furthermore, the parliament in Kerala voted in spring 2011 for a compensation payment of 216 million rupees (around €34 million) to be paid by Coca-Cola to the victims in Plachimada. However, the state president has still not signed the law, so it cannot be brought into force and a compensation tribunal cannot yet be set up.[64] It is highly doubtful that the Adivasi will ever receive such compensation. Other problems the community faces remain unheard and unresolved, in particular, a land reform that was passed in 1973 has still not progressed any farther on an administrative level.

However, it is only accurate in a very limited sense to say that the Adivasi were marginalized by opponents of globalization within the anti–Coca-Cola campaign. Such a view assumes that the Adivasi had a fixed, immobile point of view within the constantly expanding movement, and it essentializes our understanding of this group in an unacceptable way. The Adivasi were not used by intellectuals and activists; instead, they themselves came to conceive their fight against the local factory as part of a more fundamental struggle of underprivileged groups against a powerful capitalist foe. Many Adivasi and Dalits from the surrounding area supported the struggle, although they did not personally experience the effects of the plant and could not expect any advantages from it. Instead, they recognized very early on that the movement was a fundamental form of resistance against exploitation and repression, which was also perpetrated by local elites and the caste system. They also saw their role in this struggle as clearly representative of all other marginalized groups caught up in the process of globalization.[65]

When the solidarity committee was first formed and the protests won the support of other groups and NGOs, and when the regional press reported on the conflict, the movement itself and the way it was perceived still centered entirely on the issue of water. This problem was concrete and could be plainly communicated. Roles were easy to assign: the affected residents were the victims and the heroes of the resistance, whereas Coca-Cola took on the role of the powerful villain. This discourse sidelined the fundamental, far-reaching social protest by the Adivasi and the Dalits against their marginalization in Kerala. "They continued to play a crucial role within the larger movement, but they had to accept a constrained agenda."[66]

In the "Plachimada Declaration" the activists involved in the solidarity movement presented the conflict as a symbolic and internationally significant struggle because it revolved around the fundamental human right to water. The Plachimada movement was no longer led by local residents; instead, other activists groups in the Adivasi movement (Adivasi Sarakshana Samiti [ASS], Haritha Development Association, and the People's Union for Civil Liberties) wielded the greatest influence. Although local activists were involved in all the decision making, the role of most Adivasi was nevertheless reduced to taking part in sit-ins and other campaigns on the ground.[67] In this context, we can speak of a certain "professionalization" of the protest, as experienced networkers and organizers took control of the campaign when it came to dealing with the media.

However, a factor in this uncertain balance that should absolutely not be overlooked is the formidable capacity of the Adivasi for political action. They not only organized a successful local protest and held out against political and state resistance over the course of hundreds of days, they also managed with the help of their own activists to form a broad alliance to achieve their goals. It may be true that some Adivasi did not have a clear idea of what José Bové, for example, wanted to say with his speech at the World Water Conference or about the specifics of the figures discussed in his talk. But ultimately it was not that important to the affected residents whether communists, opponents of globalization, or other organizations co-opted the protest for their own political goals. Through the involvement of others, the local residents' concerns attracted greater attention and in the end this alliance was only able to achieve its aims because of the way it was framed in a discourse about globalization.

The Plachimada conflict could be worked into multiple different discourses and this meant it won the support of an incredibly diverse group of organizations. Sympathizers were able to relate to the conflict in multiple ways. In the eyes of antiglobalization groups such as Attac, the conflict was a classic case of the Global South being exploited by a multinational firm from industrialized countries, as well as a struggle against the increasing privatization of public resources such as water. For the communists, it was a class war of the proletariat against an archetypal capitalist company, and it was a case of U.S. imperialism. For environmental organizations or for fishermen from Kerala, it was about preventing water pollution. Adivasi groups saw it as part of the struggle against their marginalization in Indian society. For the general public in India, there were significant parallels between the forms of protest and the struggle for independence from the British Empire, and in this respect the movement could draw on anticolonialist discourse ("Quit India") and win the sympathy of broader groups who had less in common politically with, for example, the communists.[68]

As an increasing number of groups became involved, the conflict also expanded across different sites: from the local protests in front of the factory gates, to the arena of journalistic debate, and finally to the forum of politics and the judiciary. The original, local movement would not, by itself, have been able to broaden its reach in this way. For one thing, it would have lacked the financial resources.[69] It also did not have the legal and political experience to draw various state authorities and courts into the struggle. Without the broad support of NGOs, the protest would have had no chance of success. Conversely, the steadfast resistance of the Adivasi was the backbone of the movement, and without their daily protests that lasted for many years, the plant would never have been closed.

It should be noted that it was not the Coca-Cola Company that originally shifted the conflict into a political and thus also a legal arena. The protest movement initially caused the Panchayat to change its stance and take administrative action against the factory. In later interviews, activists attributed this to a change in strategy by party headquarters, which party representatives in the Panchayat were bound by.[70]

The big loser in the conflict was the Coca-Cola Company, which was not able to solve the crisis in India using traditional means. The corporation had tried to resolve the environmental conflict in Plachimada by using four different strategies: (1) by having attorneys assert its water rights in court, (2) by having experts provide counter-information and deny that groundwater was being endangered or contaminated, (3) by corrupting state officials and opponents, and (4) by launching an intensive PR campaign to draw attention to its own position and to undermine the arguments of its adversaries. However, these measures were futile partly because Coca-Cola did not immediately grasp the symbolic significance of the conflict and failed to understand it for a long time; the result was that all its actions ended up strengthening the position of the protestors and earning them greater popularity.

In public debate, the company was always attacked as a whole, which may certainly seem appropriate given the management structures of a multinational corporation. Nevertheless, the way Coca-Cola's manager in India acted was noticeably different from what would have been seen as an appropriate reaction in the United States or Europe to a similar environmental conflict. Instead of approaching protestors to remedy the problems, as European and American business leaders are taught to do in crisis-management seminars, the management in India categorically denied the environmental effect of the plant and tried to circumvent the ban on operations by other means—that is, corruption. Thus, even at firms that are correctly characterized as global players, a distinction should be drawn between the general company strategy and the behavior of its representatives in a particular location. The Indian branch of Coca-Cola approached its national strategy and released communications in a way that had far more in common with the behavior of other Indian companies than with its American counterpart. In this regard, the discursive framing of the struggle as a conflict between a global corporation and its local victims only partly corresponds to the reality.

However, the Coca-Cola Company showed that it had the ability to learn. After Neville Isdell took charge of the company in Atlanta, Coca-Cola adopted a new approach to public relations. Today, the company presents itself in a way that emphasizes sustainability and it engages in an active manner, designed to appeal to the public, in numerous environmental and water projects, particularly those related to improving irrigation systems. According to the company's statements, it is working more and more frequently with closed water systems and is actively championing water recycling.[71] It cannot be denied that initiatives and projects sponsored by Coca-Cola also have positive effects. Nevertheless, accusations of "greenwashing" that various environmental groups have leveled at the Coca–Cola Company are by no means unfounded, as long as the corporation fails to alter its business practices in the Global South.[72] In this respect, the soft-drink giant's voluntary environmental commitments are quite risky; PR of this sort could blow up in Coca-Cola's face if there was a potential new environmental scandal.

From a cultural sciences perspective it can be stated that this local environmental conflict would probably have taken an entirely different course if the opponent had not been Coca-Cola, a world-famous brand that attracted public attention. Coca-Cola is the world's leading soft drink manufacturer, and people on every continent consume its products. Therefore, Coca-Cola has become a symbol, together with the fast-food chain McDonald's, for the interconnectedness of the global economy—or for the expansion of the capitalist system, depending on your political perspective. If they had been up against a local Indian manufacturer without Coca-Cola's fame, it is unlikely that the Adivasi would have been able to achieve the same victory—as demonstrated by hundreds of other similar conflicts in India.

For example, no protest sprang up against a regional brewery in Meenakshipuram. The brewing facility, owned by Kerala Alcoholic Products, was only a kilometer away from Plachimada and used a comparable amount of water. The water was partly extracted from groundwater wells, and partly from a nearby canal. However, according to environmentalists' statements, neither environmental pollution nor water scarcity occurred there.[73] Students in Europe and the United States could be mobilized to take part in a boycott movement that could be easily explained and was against a highly symbolic opponent like Coca-Cola; the same could not be said of a local population's struggle against structural poverty and their fight for a land reform that had not yet been implemented.

An understanding of the importance of publicity was no longer confined to Europe or the United States; since the 1970s, an increasing number of movements in the Global South have also grasped this new context and have made use of emerging forms of communication that are becoming ever more closely interlinked. Since then, environmental conflicts of this type have increasingly transcended their own borders and have drawn in players who are not directly affected by the specific issue in question.

In this respect, the Plachimada conflict is evidence of an expanded outlook on the part of protesters, who in this case were Adivasi—a group of indigenous Eravalar and Malasar communities, who were still officially designated by Kerala's government as "primitive tribes." Since a new generation of Adivasi, exemplified, for example, by J. K. Janu, has risen to become part of India's cultural elite (thanks partly to targeted support for minorities by the federal government) and has been able to represent its communities as spokespeople, other similarly small minorities have had the chance to publicize their own concerns. "Yes, the subaltern can speak," to use a formulation from a famous essay by Gayatri Spivak.[74] Anyone coming from a Western, European or U.S.-centric perspective should not underestimate the agency of local players. In Kerala, literacy levels are over 88 percent, the Adivasi have access to radio and television, private Internet connections are becoming more common, Internet cafés can be found in all small and medium-size towns, Indian villages are also connected by phone—as long as the electricity does not cut out.

This study can thus be read as a firm commitment to write the history of apocalyptic tropes, environmental scares, and environmental crises from a perspective that is less Eurocentric (or U.S.-centric), and less focused on alleged decision makers in urban centers. Instead the issue should be analyzed "from below." The players on the ground, whether they are in Europe, the United States, or Plachimada, should receive at least as much attention as the supposedly powerful actors at the center of things. Unlike alarmist discourse, environmental conflicts take place in a specific location; however, they can cause ripples that travel far beyond the boundaries of their own immediate surroundings. Plachimada was therefore more than just a symbol.

NOTES

INTRODUCTION. THE APOCALYPTIC MOMENT

1. Rick Perlstein, *Nixonland: The Rise of a President and the Fracturing of America* (New York: Scribner, 2008), 460n. Rachel Carson, *Silent Spring* (New York: Houghton Mifflin, 1962).

2. Cf. Spencer R. Weart, *The Discovery of Global Warming* (Cambridge, MA: Harvard University Press, 2008).

3. This introduction uses alarmism and apocalypticism interchangeably, and it does not propose supertemporal criteria for alarmism: the political and moral issues at stake do not allow a consensual definition, and criteria would inevitably reflect a political agenda. For the purpose of this volume, alarmism is a type of political rhetoric that draws overwhelmingly on negative projections of future developments, usually in the form of horror scenarios that only determined action can forestall.

4. Paul Sabin, *The Bet: Paul Ehrlich, Julian Simon, and our Gamble over Earth's Future* (New Haven, CT: Yale University Press, 2013).

5. Bjørn Lomborg, *The Skeptical Environmentalist: Measuring the Real State of the World* (Cambridge: Cambridge University Press, 2001). Joseph L. Bast, Peter J. Hill, and Richard C. Rue, *Eco-Sanity: A Common-Sense Guide to Environmental-*

ism (Lanham, MD: Madison Books, 1994), 51. Pascal Bruckner, *The Fanaticism of the Apocalypse: Save the Earth, Punish Human Beings* (Cambridge: Polity, 2013), 3. Naomi Oreskes and Erik M. Conway, *Merchants of Doubt: How a Handful of Scientists Obscured the Truth on Issues from Tobacco Smoke to Global Warming* (London: Bloomsbury, 2012). Naomi Oreskes and Erik M. Conway, *The Collapse of Western Civilization: A View from the Future* (New York: Columbia University Press, 2014).

6. Oreskes and Conway, *Collapse*, 79.

7. Robinson Meyer, "Are We as Doomed as That *New York Magazine* Article Says? Why It's So Hard to Talk about the Worst Problem in the World," *Atlantic*, July 2017.

8. Times being as they are, it should come as no surprise that conspiracy theories surround environmental activism, and alarmism serves as a lightning rod. For one example, see Torsten Mann, *Rote Lügen in Grünem Gewand: Der kommunistische Hintergrund der Öko-Bewegung* (Rottenburg: Kopp, 2009).

9. Frank Kermode, *The Sense of an Ending: Studies in the Theory of Fiction* (London: Oxford University Press, 1966), 8.

10. Birgit Metzger, *"Erst stirbt der Wald, dann Du!" Das Waldsterben als deutsches Politikum (1978–1986)* (Frankfurt: Campus, 2015).

11. Jacob Darwin Hamblin, *Arming Mother Nature: The Birth of Catastrophic Environmentalism* (New York: Oxford University Press, 2013). See also John R. McNeill and Corinna R. Unger, eds., *Environmental Histories of the Cold War* (New York: Cambridge University Press, 2010).

12. Philipp Felsch, *Der lange Sommer der Theorie: Geschichte einer Revolte 1960–1990* (Munich: Beck, 2015), 156.

13. Joachim Radkau, *The Age of Ecology* (Malden, MA: Polity, 2013).

1. POWER, POLITICS, AND PROTECTING THE FOREST

1. Johann Nepomuk von Schwerz, *Beschreibung der Landwirthschaft in Westfalen und Rheinpreussen*, pt. 2 (Stuttgart: Hoffmann'sche Verlags-Buchhandlung, 1836), 136. Translations to English are the author's unless otherwise indicated.

2. Cf. Irmund Wenzel, *Ödlandentstehung und Wiederaufforstung in der Zentraleifel*, Arbeiten zur Rheinischen Landeskunde, 18 (Bonn: Dümmler, 1962). Werner Schwind, "Der Wald der Vulkaneifel in Geschichte und Gegenwart" (PhD diss., University of Göttingen, 1983).

3. Cf. Uwe Beckmann and Birgit Freese, eds., *Hölzerne Zeiten: Die unendliche*

Karriere eines Naturstoffes, Forschungsbeiträge zu Handwerk und Technik, 6 (Hagen: Westfälisches Freilichtmuseum, 1994). Joachim Radkau and Ingrid Schäfer, *Holz: Ein Naturstoff in der Technikgeschichte* (Reinbek bei Hamburg: Rowohlt, 1987).

4. Heinrich von Treitschke, *Bis zu den Karlsbader Beschlüssen*, vol. 2 of *Deutsche Geschichte im Neunzehnten Jahrhundert* (Leipzig: F.W. Händel, 1926), 273–74.

5. The debate is not only relevant to the history of the environment, it is also of central importance to our understanding of industrialization. In this scenario, the changeover from regenerative sources of energy (such as wood) to fossil fuels could be seen as a reaction to a genuine wood shortage. The debate is largely edited out of an important cultural-historical examination of the relationship of Germans to the forest: Jeffrey K. Wilson, *The German Forest: Nature, Identity, and the Contestation of a National Symbol, 1871–1914* (Toronto: University of Toronto Press, 2012).

6. Because this study focuses on public perceptions of a current or imminent threat to an elemental environmental resource, it does not go into the corresponding debates within sovereign institutions (for example, between forestry and hunting authorities and the court chamber, or between the financial and interior administrations). A more detailed account of this can be found in Bernd-Stefan Grewe, "Forst-Kultur: Die Ordnung der Wälder im 19. Jahrhundert," in *Im Schatten der Macht: Kommunikationskulturen in Politik und Verwaltung 1600–1950*, ed. Stefan Haas and Mark Hengerer, 145–70 (Frankfurt am Main: Campus, 2008).

7. Archaic German: "Abgang vnnd mangel, deß holtz, so gewißlich vnnd also bar vor augen ist, zu verhütten, vnnd die wäld in wachsung, auffgang und merung zu bringen."

8. Archaic German: "Dadurch die Höltzer vnnd Wäld, in beschwärlichen vnd schädlichen abgang dermassen gerathen, wa dem bei zeiten vnd stattlich nit begegnet, täglichen je lenger je mehr, beschwährliche vnd schädliche mängel und abgang an Holtz, auch andere fehl, vnordnungen, vns, vnsern Land vnd Leuten, schirms vnd zugewandten, auch den nachkommenden, eruolgen warden." Quoted in Waldemar Wirz, "Die Forstpolitik der südwestdeutschen Forstordnungen" (PhD diss., University of Freiburg, 1953), 26, 36.

9. Cf. Peter Blickle, "Wem gehörte der Wald? Konflikte zwischen Bauern und Obrigkeiten um Nutzungs- und Eigentumsansprüche," *Zeitschrift für Wüttembergische Landesgeschichte* 45 (1986), 167–78.

10. Cf. Heinrich Rubner, *Forstgeschichte im Zeitalter der industriellen Revolution*

(Berlin: Duncker & Humblot, 1967), 38–40; August Schwappach, *Handbuch der Forst- und Jagdgeschichte Deutschlands*, vol. 1 (Berlin: Springer, 1886), 280–84, 483–91.

11. Quoted in Johannes Brückner, *Der Wald im Feldberggebiet: Eine wald- und forstgeschichtliche Untersuchung des Südschwarzwaldes* (Bühl/Baden: Konkordia, 1970), 76.

12. Bernward Selter, "Waldnutzung und ländliche Gesellschaft: Landwirtschaftlicher 'Nährwald' und neue Holzökonomie im Sauerland des 18. und 19. Jahrhunderts," *Forschungen zur Regionalgeschichte* 13 (1995), 78, 137.

13. Jürgen Schlumbohm, "Gesetze, die nicht durchgesetzt werden: ein Strukturmerkmal des frühneuzeitlichen Staates?" *Geschichte und Gesellschaft* 23, 647–63. Cf. Schwappach, *Handbuch*, 490–91.

14. Cf. Schwappach, *Handbuch*, 280–84 ("the most important source for the study of forest and hunting conditions").

15. Cf. Radkau and Schäfer, *Holz*, 100–102. Joachim Allmann, *Der Wald in der Frühen Neuzeit: Eine mentalitäts- und sozialgeschichtliche Untersuchung am Beispiel des Pfälzer Raumes 1500–1800* (Berlin: Duncker & Humblot, 1989), 67–126, 346. Joachim Radkau, *Natur und Macht: Eine Weltgeschichte der Umwelt* (Munich: Beck, 2000), 167–72.

16. Cf. Christoph Ernst, "Forstgesetze in der Frühen Neuzeit: Zielvorgaben und Normierungsinstrumente für die Waldentwicklung in Kurtrier, dem Kröver Reich und der Hinteren Grafschaft Sponheim (Hunsrück und Eifel)," in *Policey und frühneuzeitliche Gesellschaft*, ed. Karl Härter, 341–81 (Frankfurt am Main: Klostermann, 2000). Paul Warde, "Fear of Wood Shortage and the Reality of the Woodland in Europe, c. 1450–1850," *History Workshop Journal* 62 (2006), 28–57, 44–45. Richard Hölzl, *Umkämpfte Wälder: Die Geschichte einer ökologischen Revolution in Deutschland 1760–1860* (Franfurt a.M.: Campus, 2010), 68–105.

17. Georg Ludwig Hartig, "Ist es rathsam oder nöthig die Gemeinde- und Privatwaldungen der forstpolizeilichen Aufsicht zu unterwerfen?" *Forst- und Jagdarchiv von und für Preußen* 1, no. 1 (1816), 82–91, 85.

18. Cf. Winfried Schenk, "Waldnutzung, Waldzustand und regionale Entwicklung in vorindustrieller Zeit im mittleren Deutschland," *Erdkundliches Wissen* 117 (Stuttgart: Steiner, 1996), 82–116; Christoph Ernst, *Den Wald entwickeln: Ein Politik- und Konfliktfeld in Hunsrück und Eifel im 18. Jahrhundert* (Munich: Oldenbourg, 2000), 181–92.

19. Cf. Wilhelm Mantel, "Die Einnahmen aus den bayerischen Staatswaldungen seit Ausgang des 18. Jahrhunderts: Eine geschichtlich-betriebswirtschaftliche Unter-

suchung," (PhD diss., LMU Munich, 1939), 39–40. This study only uses net forest income, and compares it with the gross income from other posts, which leads to a significant underestimate of the actual financial importance of forest revenue (pp. 7–8, 39). On the question of net calculations and gross budget, see Erika Müller, *Theorie und Praxis des Staatshaushaltsplans im 19. Jahrhundert: Am Beispiel von Preußen, Bayern, Sachsen und Württemberg* (Opladen: Westdeutscher Verlag, 1989).

20. Quoted by Ernst, *Den Wald entwickeln*, 214–15.

21. Ernst, *Den Wald entwickeln*, 223.

22. Ernst, *Den Wald entwickeln*, 242–43, 329–30. Imperial cities such as Nuremberg were hardly behind the territorial lords in their politics of the forest: Herbert May and Markus Rodenberg, eds., *Der Reichswald: Holz für Nürnberg und seine Dörfer* (Bad Windsheim/Lauf an der Pegnitz: Fränkisches Freilandmuseum, 2013).

23. Selter, "Waldnutzung," 72–73.

24. Quoted in Wolfgang von Hippel, *Auswanderung aus Südwestdeutschland: Studien zur württembergischen Auswanderungspolitik im 18. und 19. Jahrhundert* (Stuttgart: Klett-Cotta, 1984), 300–305, 303–4.

25. Quoted in Georg Fertig, *Lokales Leben, atlantische Welt: Die Entscheidung zur Auswanderung vom Rhein nach Nordamerika im 18. Jahrhundert* (Osnabrück: Rasch, 2000), 193. For more on wood shortages as a reason for emigration, see Sigrid Faltin, *Die Auswanderung aus der Pfalz nach Nordamerika im 19. Jahrhundert: Unter besonderer Berücksichtigung des Landkommissariates Bergzabern* (Franfurt a.M.: P. Lang, 1986), 87–90.

26. The conflict between rulers and subjects was played out in various, often interrelated, arenas. See Bernd-Stefan Grewe, "Streit um den Wald—ein Ressourcenkonflikt? Das Konfliktfeld Wald in der vorindustriellen Zeit (ca. 1500–1850)," *Geschichte in Wissenschaft und Unterricht* 63 (2012), 551–66. See also the local studies: Hölzl, *Umkämpfte Wälder*. Niels Grüne, *Dorfgesellschaft—Konflikterfahrung—Partizipationskultur: Sozialer Wandel und politische Kommunikation in Landgemeinden der badischen Rheinpfalz (1720–1850)* (Stuttgart: Lucius & Lucius, 2011). On legal disputes over the forest see Stefan von Below and Stefan Breit, *Wald: Von der Gottesgabe zum Privateigentum: Gerichtliche Konflikte zwischen Landesherren und Untertanen um den Wald in der frühen Neuzeit* (Stuttgart: Lucius & Lucius, 1998). Jonathan Sperber, "Angenommene, vorgetäuschte und eigentliche Normenkonflikte bei der Waldnutzung im 19. Jahrhundert," *Historische Zeitschrift* 290 (2010), 681–702. Manfred Hörner, "Wald und Wald-

nutzung als Gegenstand von Reichskammergerichtsprozessen," *Archive in Bayern 7* (2012), 311–25.

27. On this case and for the quotes, see Schenk, "Waldnutzung," 210.

28. Cf. Selter, "Waldnutzung," 218–32.

29. Ingrid Schäfer, "'Gewerbehierarchie'—Instrument der Brennstoffpolitik im 18. Jahrhundert: Sozial- und technikgeschichtliche Aspekte zur Holzversorgung in den Fürstentümern Lippe-Detmold und Nassau-Dillenburg, *Scripta Mercaturae* 17 (1983), 63–90, 70–72. Cf. Ingrid Schäfer, *"Ein Gespenst geht um": Politik mit der Holznot in Lippe 1750–1850. Eine Regionalstudie zur Wald- und Technikgeschichte* (Detmold: Naturwissenschaftlicher und Historischer Verein für das Land Lippe, 1992). Ingrid Schäfer, *Privatwald in Lippe: Natur und Ökonomie zwischen 1750 und 1950* (Bielefeld: Verlag für Regionalgeschichte, 1998).

30. Numerous examples are cited (albeit with the opposite intent) in Uwe Eduard Schmidt, "Das Problem der Ressourcenknappheit dargestellt am Beispiel der Waldressourcenknappheit in Deutschland im 18. und 19. Jahrhundert: Eine historisch-politische Analyse" (Habilitation thesis, LMU Munich, 1997), 32.

31. This source also appears in Burkhard Dietz, *Wirtschaftliches Wachstum und Holzmangel im bergisch-märkischen Gewerberaum vor der Industrialisierung* ([n.p.]: 1997), 167.

32. On the role of the state in creating a free(er) market, see Margit Grabas, "Krisenbewältigung oder Modernisierungsblockade? Die Rolle des Staates bei der Überwindung des 'Holzenergiemangels' zu Beginn der Industriellen Revolution in Deutschland," *Jahrbuch für Europäische Verwaltungsgeschichte 7* (1995), 43–75. On commercial forest politics, see August Bernhardt, *Geschichte des Waldeigentums, der Waldwirtschaft und Forstwissenschaft in Deutschland*, vol. 2 (Berlin: Springer, 1874), 66–70.

33. See also the corresponding article in the contemporary reference book, Johann Georg Krünitz, ed., *Oeconomische Encyclopädie*, 242 vols. (Berlin: Pauli, 1773–1858), vols. 24–32 titled *Oekonomische Encyklopädie*, from vol. 33 titled *Oekonomisch-technologische Encyklopädie*. J. S. Ersch and J. G. Gruber, eds., *Allgemeine Encyklopädie der Wissenschaften und Künste*, 61 vols. (Leipzig: Gleditsch, 1818–89).

34. Studies by Rolf-Jürgen Gleitsmann are the seminal works on the wood-saving literature. See Gleitsmann, "Rohstoffmangel und Lösungsstrategien: Das Problem vorindustrieller Holzknappheit," in *Technologie und Politik: Das Magazin zur Wachstumskrise*, ed. Freimut Duwe, 104–54, (Reinbek bei Hamburg: Rowohlt, 1980), 121–26. Gleitsmann, "Erfinderprivilegien auf holzsparende Technologien

im 16. und frühen 17. Jahrhundert," *Technikgeschichte* 52 (1985), 217–32. Gleitsmann, "Und immer starben die Wälder: Ökosystem Wald, Waldnutzung und
Energiewirtschaft in der Geschichte," *Mensch und Umwelt in der Geschichte*,
ed. Jörg Calließ et al., Geschichtsdidaktik vol. 5, 175–204 (Pfaffenweiler: Centaurus, 1989). See also Eckart Reidegeld, "Nachhaltige Forstwirtschaft und
Holzsparkunst: Frühe Formen des Umgangs mit Ressourcenknappheit," *Leviathan* 42 (2014), 433–62.

35. Carl Fischer, quoted in Eberhard Elbs, "'Holznot' und 'Holzsparkunst': Zur Krise
des Waldes im 18. Jahrhundert," *Schwäbische Heimat* 38 (1987), 297–306, 303. The
most important work on domestic fires is Alfred Faber, *Entwicklungsstufen der
häuslichen Heizung* (Munich: Oldenbourg, 1957).

36. A seminal work on the development of the iron industry is Rainer Fremdling,
*Technologischer Wandel und internationaler Handel im 18. und 19. Jahrhundert:
Die Eisenindustrien in Großbritannien, Belgien, Frankreich und Deutschland*
(Berlin: Duncker & Humblot, 1986).

37. See the following examples, selected according to their regional origin: Georg
Ludwig Hartig, *Beweiß, daß durch die Aufzucht der weißblühenden Acacie
schon wirklich entstandenem Brennholzmangel nicht abgeholfen werden kann:
Nebst einem Vorschlag, auf welche Art dieser große Zweck viel sicherer zu
erreichen seyn möchte* (n.p., 1798) (Nassau-Dillenburg, later the head of the
Prussian forest administration). Wilhelm Gottfried von Moser, *Gedanken
zum Holzmangel* ([n.p.]: 1762) (Württemberg). K. Papius, *Die Holznoth und
die Staatsforste* (Munich: Lindauer, 1840) (Bavaria). Wilhelm Leopold Pfeil,
*Ueber die Ursachen des schlechten Zustandes der Forsten und die allein mögli-
chen Mittel, ihn zu verbessern, mit besonderer Rücksicht auf die preußischen
Staaten* (Züllichau/Freistadt: Darnmann, 1816). J. H. Rutisch, *Versuch einer
Beantwortung der beyden Fragen: sind die Klagen über den Mangel und die
Theurung des Holzes in Sachsen begründet? und wie ist selbigen abzuhelfen?*
(n.p.: 1799). Carl August Scheidt, "Kurze Beschreibung über einige Ursa-
chen des allgemein werdenden Holzmangels in Deutschland und über die
Mittel, demselben abzuhelfen," *Abhandlungen der Churfürstlich-Bairischen
Akademie* 9 (n.p. 1775). Friedrich Ludwig v. Witzleben, *Abhandlung über ein-
ige noch nicht genug erkannte und beherzigte Ursachen des Holzmangels . . .*
(Frankfurt a Main: Herrmann, 1800) (Nassau-Dillenburg, and the Langravi-
ate of Hesse).

38. Essentially, this was a case of suppressing competing uses of the forest, see Bernd
Fuhrmann, "Holzversorgung, Waldentwicklung, Umweltveränderungen und

wirtschaftliche Tendenzen in Spätmittelalter und beginnender Neuzeit," *Viertel-jahrschrift für Sozial- und Wirtschaftsgeschichte* 100 (2013), 311–27.

39. On sovereign interests in control and surveillance, see James C. Scott, *Seeing Like a State: How Certain Schemes to Improve the Human Condition Have Failed* (New Haven, CT: Yale University Press, 1998), 11–52.

40. On the development of forest science and forestry since the Middle Ages see Bernhardt, *Geschichte des Waldeigentums.* Schwappach, *Handbuch.* Rubner, *Forstgeschichte.* Karl Hasel, *Forstgeschichte: Ein Grundriß für Studium und Praxis* (Hamburg/Berlin: Parey, 1985). Kurt Mantel, *Wald und Forst in der Geschichte: Ein Lehr- und Handbuch* (Hannover: Schaper, 1990).

41. Mantel, *Wald und Forst,* 321. Cf. Josef Köstler," Wald und Forst in der deutschen Geschichtsforschung," *Historische Zeitschrift* 155 (1937), 461–74, 469. Gertrud Schröder-Lembke, "Waldzerstörung und Walderneuerung in Deutschland in der vorindustriellen Zeit," *Zeitschrift für Agrargeschichte und Agrarsoziologie* 35 (1987), 120–37. This view of forestry as a beneficial form of forest conservation and development politics is still represented by Werner Rösener, "Der Wald als Wirtschaftsfaktor und Konfliktfeld in der Gesellschaft des Hoch- und Spätmit-telalters," *Zeitschrift für Agrargeschichte und Agrarsoziologie* 55 (2007), 14–31, 31. See also Andrea Germer, "Mensch und Wald im Hildesheimer Raum vom 16. bis 19. Jahrhundert: Ein Beitrag zur Umweltgeschichte," *Hildesheimer Jahrbuch für Stadt und Stift Hildesheim* 79 (2008), 49–79.

42. Cf. Schenk, "Waldnutzung." Selter, "Waldnutzung." Ernst, *Den Wald entwickeln.* Elisabeth Weinberger, *Waldnutzung und Waldgewerbe in Altbayern im 18. und beginnenden 19. Jahrhundert* (Stuttgart: Steiner, 2001). Johanna R. Regnath, *Das Schwein im Wald: Vormoderne Schweinehaltung zwischen Herrschaftsstrukturen, ständischer Ordnung und Subsistenzökonomie* (Ostfildern: Thorbecke, 2008). Regnath, "Die Schweinemast im Schönbuch: Eine spätmittelalterliche und früh-neuzeitliche Waldnutzungsform im Spannungsfeld von Territorialpolitik und Subsistenzökonomie," in *Landnutzung und Landschaftsentwicklung im deutschen Südwesten: Zur Umweltgeschichte im späten Mittelalter und in der frühen Neuzeit,* ed. Sönke Lorenz, 179–97 (Stuttgart: Steiner, 2009). Martin Stuber and Matthias Bürgi, *Hüeterbueb und Heitisträhl: Traditionelle Formen der Waldnutzung in der Schweiz 1800 bis 2000* (Bern: Haupt, 2011).

43. Main State Archive, Koblenz (LHAK), Inventory 442–4475 additional uses. Let-ter from forest superintendent Coupette to the district government of Trier, May 11, 1830.

44. Cf. Grewe, *Der versperrte Wald.*

45. LHAK Inventory 442 District government of Trier—10471 Generalia Forstbetrieb.

46. LHAK Inventory 403–9455 Bildung von Forstschutz- und Holzersparungsvereinen.

47. This information originates from Dirk Blasius, *Bürgerliche Gesellschaft und Kriminalität. Zur Sozialgeschichte Preußens im Vormärz* (Göttingen: Vandenhoeck & Ruprecht, 1976). Important publications on contraventions of the forest laws include: Josef Mooser, "'Furcht bewahrt das Holz': Holzdiebstahl und sozialer Konflikt in der ländlichen Gesellschaft 1800–1850 an westfälischen Beispielen," in *Räuber, Volk und Obrigkeit: Studien zur Geschichte der Kriminalität in Deutschland seit dem 18. Jahrhundert,* ed. Heinz Reif, 43–99 (Frankfurt am Main: Suhrkamp, 1984); Reiner Prass, "Verbotenes Weiden und Holzdiebstahl: Ländliche Forstfrevel am südlichen Harzrand im späten 18. und frühen 19. Jahrhundert," *Archiv für Sozialgeschichte* 36 (1996), 51–68; Bernd-Stefan Grewe, "'Darum treibt hier Not und Verzweiflung zum Holzfrevel': Ein Beitrag zur Sozial-, Wirtschafts- und Umweltgeschichte der Pfalz 1816–1860," *Mitteilungen des historischen Vereins der Pfalz* 94 (1996), 271–95; Grewe, "Das wirtschaftliche Gewicht der Kriminalität: Grenzen und Möglichkeiten statistischer Verfahren in der Forstgeschichte am Beispiel der Forstfrevel in der bayerischen Pfalz 1820–1860," in "*Lange Reihen*" *zur Erforschung von Waldzuständen und Waldentwicklungen,* ed. Winfried Schenk, 231–50 (Tübingen: Geographisches Institut der Universität Tübingen, 1999); and Grüne, *Dorfgesellschaft,* 288–91.

48. The most famous discussion of the social basis for wood theft in the region was by the young Karl Marx in 1842: Karl Marx, "Debatten über das Holzdiebstahlsgesetz: Von einem Rheinländer," in Karl Marx and Friedrich Engels, *Werke,* vol. 1 (Berlin (GDR): Dietz, 1976), 109–47. (*Rheinische Zeitung* 298, 300, 303, 305, 307 [October 25–November 3, 1842]).

49. This shortage often concerned particular types and qualities of wood. For more on early modern Württemberg, see Paul Warde, *Ecology, Economy and State Formation in Early Modern Germany* (Cambridge: Cambridge University Press, 2006), 224–79.

50. Joachim Radkau, *Die Ära der Ökologie: Eine Weltgeschichte* (Munich: Beck, 2011), 45.

51. Werner Sombart, *Der moderne Kapitalismus: Historisch systematische Darstellung des gesamteuropäischen Wirtschaftslebens von seinen Anfängen bis zur Gegenwart,* vol. 2, *Das europäische Wirtschaftsleben im Zeitalter des Frühkapitalismus,*

second half of volume (repr. Berlin: Duncker & Humblot; 2nd ed, 1916), 1137, 1153.

52. In addition to the previously cited works on forest history by Bernhardt, Schwappach, Hasel, Mantel, and Schmidt, see Rolf Peter Sieferle, *Der unterirdische Wald. Energiekrise und Industrielle Revolution*, Die Sozialverträglichkeit von Energiesystemen, 2 (Munich: Beck, 1982). Rolf-Jürgen Gleitsmann, "Der Einfluß der Montanwirtschaft auf die Waldentwicklung Mitteleuropas: Stand und Aufgaben der Forschung," in *Montanwirtschaft Mitteleuropas vom 12. bis 17. Jahrhundert: Stand, Wege und Aufgaben der Forschung* (Bochum: Vereinigung der Freunde von Kunst und Kultur im Bergbau, 1984), 24–39. Jean-Claude Debeir et al., *Prometheus auf der Titanic: Geschichte der Energiesysteme* (Frankfurt am Main: Campus, 1989; French ed.: 1986), 142–72 (the energy revolution and European industrialization). Christian Hünemörder, "Umweltschädigungen in historischer Sicht—Beispiel Mitteleuropa," in *Technik und Natur*, ed. Werner Nachtigall and Charlotte Schönbeck, Technik und Kultur, 6, 321–43 (Düsseldorf: VDI, 1994), 330.

53. On these scholarly positions, see Radkau and Schäfer, *Holz*. Joachim Radkau, "Zur angeblichen Energiekrise des 18. Jahrhunderts: Revisionistische Betrachtungen über die 'Holznot,'" *Vierteljahrschrift für Sozial- und Wirtschaftsgeschichte* 73 (1986), 1–37. Schäfer, *Ein Gespenst geht um*. Allmann, *Der Wald in der frühen Neuzeit*.

54. Cf. Ernst, *Den Wald entwickeln*, 325–40. A depiction that is less critical of the sources can be found in Hans Heinrich Vangerow, "Gedanken zur 'Holznot' in der oberen Pfalz um die Mitte des 18. Jahrhunderts," in *Forum Forstgeschichte: Ergebnisse des Arbeitskreises Forstgeschichte in Bayern 2005–2007*, ed. Egon Gundermann, 96–110 (Munich: LMU Munich, 2007).

55. Cf. Grewe, *Der versperrte Wald*. On the development of the forest into forestry categories, see Matthias Bürgi, *Waldentwicklung im 19. und 20. Jahrhundert: Veränderungen in der Nutzung und Bewirtschaftung des Waldes und seiner Eigenschaften als Habitat am Beispiel der öffentlichen Waldungen im Züricher Unter- und Weinland* (PhD diss., ETH Zurich, 1998).

56. This study does not look at deforestation caused by acts of war and the subsequent damage to wood supplies. See John R. McNeill, "Woods and Warfare in World History," *Environmental History* 3 (2004), 388–410. Ralf Faber, "Der Lippische Wald während und nach dem Dreißigjährigen Krieg," in *Das historische Erbe in der Region: Festschrift für Detlev Hellfaier*, ed. Axel Halle, 253–61 (Bielefeld: Aisthesis, 2013). Axel Bader, *Wald und Krieg: Wie sich in Kriegs- und Krisen-*

zeiten die Waldbewirtschaftung veränderte: Die deutsche Forstwirtschaft im Ersten Weltkrieg (Göttingen: Universitätsverlag, 2011).

57. Bernhardt, *Waldeigentum*, 2:100–101.

58. On the question of wood transport and how its effects shaped different areas, see Martin Knoll, "Von der prekären Effizienz des Wassers: Die Flüsse Donau und Regen als Transportwege der städtischen Holzversorgung Regensburgs im 18. und 19. Jh." *Saeculum* 58, no. 1 (2007), 33–58. Katja Hürlimann, "Raumprägende Wirkungen der Holznutzung im 18. und 19. Jahrhundert," *Schweizerische Zeitschrift für Wirtschafts- und Sozialgeschichte* 25 (2010), 179–91. Christian Lotz views the expansion of timber frontiers as a history of international integration. See Christian Lotz, "Expanding the Space for Future Resource Management: Explorations of the Timber Frontier in Northern Europe and the Rescaling of Sustainability during the 19th Century," *Environment and History* 21 (2015), 257–79.

59. Cf. Ernst, *Den Wald entwickeln*, 331. Radkau and Schäfer, *Holz*, 65. On the conflicting goals in regard to using the forest for hunting or for wood, see Martin Knoll, *Umwelt—Herrschaft—Gesellschaft: Die landesherrliche Jagd Kurbayerns im 18. Jahrhundert* (St. Katharinen: Scripta Mercaturae, 2004). Christian Kruse, ed., *WaldGeschichten: Forst und Jagd in Bayern 811—2011* (Munich: Bayerisches Hauptstaatsarchiv, 2011). Janina Wirth, "Der Nürnberger Reichswald zwischen Nutzung und Repräsentation: Mit einem Schwerpunkt auf der Jagd," in *Křivoklát—Pürglitz: Jagd, Wald, Herrscherrepräsentation*, ed. Jiří Fakjt et al., 313–54 (Ostfildern: Thorbecke, 2014).

60. Cf. Joachim Radkau, "Holzverknappung und Krisenbewußtsein im 18. Jahrhundert," *Geschichte und Gesellschaft* 9 (1983), 513–43. He writes that "there was a wish to make wood a scarce commodity" (515–16). Cf. Grewe, *Der versperrte Wald*.

61. Cf. Mantel, *Wald und Forst in der Geschichte*, 474. A particularly harsh version of this criticism can be found in Wilhelm Bode and Martin von Hohnhorst, *Waldwende: Vom Försterwald zum Naturwald* (Munich: Beck, 1994); Peter-M. Steinsiek, "Determinanten der Waldentwicklung im Westharz (16.–18. Jahrhundert)," *Niedersächsisches Jahrbuch für Landesgeschichte* 80 (2008), 117–40, 132–37; Reinhard Mosandl, "Geschichte der Wälder in Mitteleuropa im letzten Jahrtausend: Aktuelle Beiträge zum Verhältnis der historischen Entwicklung," in *Beiträge zum Göttinger Umwelthistorischen Kolloquium 2008—2009*, ed.Bernd Herrmann (Göttingen: Universitätsverlag, 2009), 91–114, 110.

2. GRASSROOTS APOCALYPTICISM

1. *LIFE Magazine* 68, no. 3 (January 30, 1970), 22.

2. *Time Magazine* 95, no. 5 (February 2, 1970), 59.

3. David Dietz, "Says Pollution Could Erase Human Race in 30 Years," *Cleveland Press*, November 17, 1970.

4. "Pollution Seen Destroying Democracy by Year 2000," *Plain Dealer* (Cleveland), January 19, 1970.

5. J. Clarence Davies III, *The Politics of Pollution* (New York: Pegasus, 1970), 193. See also chapter 3 in this volume, by Patrick Kupper and Elke Seefried.

6. Huntington Library, Manuscripts Department, San Marino, California, Kenneth Hahn Collection Box 224 Folder 5, Joint Meeting of the Scientific and Emergency Action Committees, September 15, 1970, Agenda, Point 4 Item a. For Libby's stance on nuclear matters, see Allan M. Winkler, *Life under a Cloud: American Anxiety about the Atom* (Urbana: University of Illinois Press, 1999), 104.

7. John McCormick, "The Prophets of Doom," in *Reclaiming Paradise: The Global Environmental Movement* (Bloomington: Indiana University Press, 1991), ch. 4.

8. Jon Gabriel, "13 Worst Predictions Made on Earth Day, 1970," April 22, 2013, (http://www.freedomworks.org/content/13-worst-predictions-made-earth -day-1970; accessed January 14, 2018).

9. Cf. David Stradling, *Smokestacks and Progressives: Environmentalists, Engineers and Air Quality in America, 1881–1950* (Baltimore: Johns Hopkins University Press, 1999); Frank Uekötter, *The Age of Smoke: Environmental Policy in Germany and the United States, 1880–1970* (Pittsburgh: University of Pittsburgh Press, 2009).

10. James C. Williams, *Energy and the Making of Modern California* (Akron: University of Akron Press, 1997), p. 364.

11. Lucius H. Cannon, *Smoke Abatement: A Study of the Police Power as Embodied in Laws, Ordinances and Court Decisions* (St. Louis: St. Louis Public Library, 1924), 258.

12. John E. Baur, *The Health Seekers of Southern California, 1870–1900* (San Marino, CA: Huntington Library, 1959).

13. John Anson Ford Collection, Huntington Library, San Marino, California, Box 25 B III 5 a aa, Mayor's Conference on Control of Smoke and Fumes. Reporter's Transcript of Proceedings on May 8, 1946, p. 6.

14. For an overview on the history of Los Angeles smog, see Marvin Brienes, "The Fight against Smog in Los Angeles, 1943–1957" (PhD diss., University of Cali-

fornia Davis, 1975); James E. Krier and Edmund Ursin, *Pollution and Policy: A Case Essay on California and Federal Experience with Motor Vehicle Air Pollution 1940–1975* (Berkeley: University of California Press, 1977); Scott Hamilton Dewey, *Don't Breathe the Air: Air Pollution and U.S. Environmental Politics, 1945–1970* (College Station: Texas A&M University Press, 2000).

15. University of California, Los Angeles, Department of Special Collections, Charles E. Young Research Library, Collection 1108 (Richard Richards Papers) Box 38 Folder 1, Citizens Anti-Smog Action Committee, Francis H. Packard, Chairman, Pasadena, October 13, 1955, p. 1

16. Richard Richards Papers Box 39 Folder 1, Henry Gurewitz to Senator Richard Richards, undated (ca. January 1959).

17. Kenneth Hahn Collection Box 266 Folder 3b, Clean Air for California. Second Report of the California State Department of Public Health, March 1956, p. 10.

18. Quoted from Helen B. Shaffer, "Poisoned Air," *Editorial Research Reports* 1955, 239–56, 248.

19. These disasters were mentioned in Richard Richards Papers Box 37 Folder 4, State of California, Department of Public Health, Clean Air for California. Initial Report of the Air Pollution Study Project, March 1955, p. 35.

20. Benoit Nemery, Peter H. M. Hoet, and Abderrahim Nemmar, "The Meuse Valley Fog of 1930: An Air Pollution Disaster," *Lancet* 357 (2001), 704–8, 708.

21. Franklin D. Roosevelt Presidential Library, Hyde Park, New York, James Roosevelt Papers Container 271 Folder "Smog, Folder 1," A. A. Golden, "The Smog Menace," 2.

22. Yale University, Manuscripts and Archives, Sterling Memorial Library, New Haven, Connecticut, Edwin Richard Weinerman Papers Box 63 Folder 125, "The Problem of Smog," 17th KPFA Medical Care Broadcast, April 14, 1956, p. 5.

23. L. McCabe and G. Clayton, "Air Pollution by Hydrogen Sulphide in Poza Rica, Mexico," *Archives of Industrial Hygiene and Occupational Medicine* 6 (1952), 199.

24. Philip Drinker, "Deaths during the Severe Fog in London and Environs, Dec. 5 to 9, 1952," *A.M.A. Archives of Industrial Hygiene and Occupational Medicine* 7 (1953), 275–76; John A. Scott, "Fog and Deaths in London, December 1952," *Public Health Reports* 68 (1953), 474–79.

25. For perspectives on the Donora disaster, see Lynne Page Snyder, "'The Death-Dealing Smog over Donora, Pennsylvania': Industrial Air Pollution, Public Health, and Federal Policy, 1915–1963" (PhD diss., University of Pennsylvania, 1994); Devra Davis, *When Smoke Ran Like Water: Tales of Environmental Deception and the Battle against Pollution* (New York: Basic Books, 2002), 5–30.

26. Reports usually mention twenty victims, though at least fifty more people died in the month after the episode. Even the local monument to the disaster ignores these additional casualties. (Davis, *When Smoke Ran Like Water*, 27, 29.)

27. This point is made with particular vigor in George A. Gonzalez, "Urban Growth and the Politics of Air Pollution: The Establishment of California's Automobile Emission Standards," *Polity* 35 (2002), 213–36.

28. Brienes, "Fight against Smog," 117.

29. John Anson Ford Collection Box 26 B III 5 a ff Folder 1, Simon to Ford, April 19, 1950.

30. Kenneth Hahn Collection Box 266 Folder 1a, Reed to Hahn, October 14, 1953.

31. Washington University Archives, St. Louis, Raymond R. Tucker Smoke Abatement Collection Series 3 Box 1 Folder "Los Angeles (Letter from Individuals)," Stewart to Tucker, December 13, 1946.

32. Kenneth Hahn Collection Box 266 Folder 1a, Arnow to Hahn, October 8, 1953.

33. Cincinnati Medical Heritage Center, University of Cincinnati Medical Center, Robert A. Kehoe Archives Box 46 Grouping 13, Pottenger to Kehoe, December 15, 1950.

34. Robert A. Kehoe Archives Box 46 Grouping 13, Pottenger to Kehoe, December 15, 1950.

35. Robert A. Divine, *Blowing in the Wind: The Nuclear Test Ban Debate 1954–1960* (New York: Oxford University Press, 1978).

36. Rachel Carson, *Silent Spring* (Twenty-fifth Anniversary ed.) Boston: Houghton Mifflin, 1987), 1.

37. Kenneth Hahn Collection Box 266 Folder 1a, Hahn to Larson, October 2, 1953.

38. Uekötter, *Age of Smoke*, 204.

39. Adam W. Rome, "Coming to Terms with Pollution: The Language of Environmental Reform, 1865–1915," *Environmental History* 1, no. 3 (July 1996), 16.

40. Metro Clean Air Committee Records, Minnesota Historical Society, St. Paul, Box 1 Folder "General" no. 1, PCA Hearing—Statement on SO and Particulates, March 25, 1970, p. 1.

41. Hazel Erskine, "The Polls: Pollution and Its Costs," *Public Opinion Quarterly* 36 (1972), 121.

42. Archives of Industrial Society, University of Pittsburgh, 80:7 Box 1 FF 1, County of Allegheny, Bureau of Smoke Control, A Review of Program, June 18, 1951, p. 2.

43. Historical Society of Western Pennsylvania Archives, Pittsburgh Mss 285 File "Smoke Control—General," Edward L. Stockton, "Air and Water Pollution,"

paper presented at the Western Regional PPHA Meeting at Mellon Institute, Pittsburgh, Pennsylvania, April 19, 1967, p. 7.

44. Kenneth Hahn Collection Box 266 Folder 1a, Hahn to Larson, November 19, 1953.

45. United States Senate, Committee on Public Works, "A Study of Pollution—Air," Staff Report, 88th Congress, 1st Session, September 1963 (Washington, DC: Government Printing Office, 1963), 33.

46. J. M. Dallavalle, "Methods of Control of Atmospheric Pollution," *Air Conditioning, Heating and Ventilation* 53, no. 9 (September 1956), 75–78, 75.

47. James J. Hanks and Harold D. Kube, "Industry Action to Combat Pollution," *Harvard Business Review* 44, no. 5 (September 1966), 49–62, 62.

48. Temple University, Philadelphia, Urban Archives, Greater Philadelphia Movement Records Box 12 Folder "Air Pollution," *Let's Clear the Air* 1 no. 10 (June 1968), 1.

49. Minnesota Historical Society, St. Paul, Metro Clean Air Committee Records Box 2 Folder "Unions," American Federation of Labor and Congress of Industrial Organizations, Man and his Environment, January 1969, p. 7.

50. Western Reserve Historical Society, Cleveland, Carl B. Stokes Papers Container 11 Folder 171, 3rd Grade Creative Writing Earth Patrol Project, St. John Lutheran School.

51. National Archives of the United States of America RG 90 A 1 Entry 11 Box 42 Folder "OCC—'C,'" The Conservation Foundation, CF Commentary, July 15, 1966, p. 1.

52. Quoted in W. A. Raleigh Jr., "Coal's Stake in Reducing Smog," *Coal Age* 61, no. 10 (October 1956), 54–59, 55.

53. United States Senate, "Study of Pollution," 51.

54. The Bancroft Library, University of California, Berkeley, MSS 71/103 c Sierra Club Records Container 113 Folder 12, Luten to McCloskey (1967).

55. "No Laughing Matter: The Cartoonist Focuses on Air Pollution," Public Health Service Publication no. 1561 (Washington, DC: Government Printing Office, 1966), 46.

56. Cf. Samuel P. Hays, *Beauty, Health, and Permanence: Environmental Politics in the United States, 1955–1986* (Cambridge: Cambridge University Press, 1987); J. Brooks Flippen, *Nixon and the Environment* (Albuquerque: University of New Mexico Press, 2000); Adam Rome, *The Genius of Earth Day: How a 1970 Teach-in Unexpectedly Made the First Green Generation* (New York: Hill and Wang, 2003).

57. Minnesota Historical Society, St. Paul, Steve J. Gadler Papers Box 1 Folder "Horn, Charles Lilley," Kaupanger to Bethlehem Steel Corporation, April 22, 1970.

58. *Playboy* 18, no. 8 (August 1971), 156.

59. Robert Moses, "Bomb Shelters, Arks and Ecology," *National Review* 22, no. 35 (September 8, 1970), 938–42, 939.

60. *Time Magazine* 96, no. 5 (August 3, 1970), 42.

61. Historical Society of Western Pennsylvania Archives, Pittsburgh, GASP Records Mss 43 Series II Box 3 Folder 11, How To Testify At A Public Hearing (undated). On the background of Pittsburgh antipollution activism, see James Longhurst, *Citizen Environmentalists* (Medford, MA: Tufts University Press, 2010).

62. Sierra Club Records Container 131 Folder 3, People's Lobby, Inc., Initiative Measure to Be Submitted Directly to the Electors (1969).

63. Sierra Club Records Container 94 Folder 1, Ela to Dedrick, December 15, 1969, p. 1.

64. Sierra Club Records Container 94 Folder 1, Ela to Dedrick, December 15, 1969, p. 2.

65. Sierra Club Records Container 122 Folder 3, Murphy to Biddle, September 9, 1969, and Maharg and Fuller to Los Angeles County Board of Supervisors, November 17, 1969, p. 1.

66. Sierra Club Records Container 94 Folder 1, Ela to Dedrick, December 15, 1969, p. 2. For an overview of the Sierra Club's history, see Michael P. Cohen, *The History of the Sierra Club 1892–1970* (San Francisco: Sierra Club Books, 1988).

67. Oregon Historical Society, Portland, Oregon Environmental Council Papers Box 19 Folder 10, Coalition for Clean Air, Oregon/Washington, Minutes of the Executive Committee Meeting, December 14, 1972.

68. World Health Organization, "7 Million Premature Deaths Annually Linked to Air Pollution," March 25, 2014 (http://www.who.int/mediacentre/news/releases/2014/air-pollution/en/; accessed January 14, 2018).

69. Jeanne Kuebler, "Air Contamination," *Editorial Research Reports 1964*, 3–19, 7.

3. "A COMPUTER'S VISION OF DOOMSDAY"

1. Donella H. Meadows, Dennis L. Meadows, Jørgen Randers, and William W. Behrens III, *The Limits to Growth: A Report for the Club of Rome's Project on the Predicament of Mankind* (New York: Universe Books, 1972).

2. Aurelio Peccei, *The Chasm Ahead* (London: Collier Macmillan, 1969), 145. For Peccei's view of the world and his proximity to the "evolutionary humanism"

of Julian Huxley, see Elke Seefried, *Zukünfte: Aufstieg und Krise der Zukunfts-forschung 1945–1980* (Berlin: De Gruyter Oldenbourg, 2015), 242–47. See also Élodie Vieille Blanchard, "Technoscientific Cornucopian Futures versus Dooms-day Futures: The World Models and The Limits to Growth," in *The Struggle for the Long-Term in Transnational Science and Politics*, ed. Jenny Anderson and Eglė Rindzevičiūtė, 92–114 (New York: Routledge, 2015), 97–99. For interview-based material, see Peter Moll, *From Scarcity to Sustainability: Futures Studies and the Environment: The Role of the Club of Rome* (Frankfurt am Main: Peter Lang, 1991), 49–63. See also Luigi Piccioni, "Forty Years Later: The Reception of the Limits to Growth in Italy 1971–74," *I quaderni di altronovecento* 2 (2012). For autobiogra-phy, see Aurelio Peccei, *The Human Quality* (Oxford: Pergamon Press, 1977).

3. Aurelio Peccei, "The Challenge of the 1970s for the World of Today" (1965), in *The Club of Rome*, ed. Pentti Malaska and Matti Vapaavuori, 10–20 (Turku: Finn-ish Society for Futures Studies, 1984).

4. Peccei, *Chasm*, 243. Cf. Seefried, *Zukünfte*, 241–42, 244–45.

5. Cf. Seefried, *Zukünfte*, 235–41. Matthias Schmelzer, "The Crisis before the Cri-sis: The 'Problems of Modern Society' and the OECD, 1968–1974," in *European Review of History* 6, no. 19 (2012), 999–1020. Matthias Schmelzer, "'Born in the Corridors of the OECD': The Forgotten Origins of the Club of Rome, Transna-tional Networks, and the 1970s in Global History," *Journal of Global History* 12, no. 1 (2017), 26–48. The OECD report was published as Erich Jantsch, *Techno-logical Forecasting in Perspective: A Framework for Technological Forecasting, Its Techniques and Organisation* (Paris: OECD, 1967).

6. Cf. Alexander King, "The Club of Rome and Its Policy Impact," in *Knowledge and Power in the Global Society*, ed. William M. Evan, 205–24 (Beverly Hills: Sage, 1981), 206–7. Peccei, *Human Quality*, 62–64.

7. Club of Rome, "The Predicament of Mankind: Quest for Structured Responses to Growing World-Wide Complexities and Uncertainties: A Proposal" (1970), in OECD Archives, Folder 218055, Supplement 218062.

8. See Moll, *Scarcity*, 49–92. See also Jürgen Streich, *30 Jahre Club of Rome: Anspruch, Kritik, Zukunft* (Basel: Birkhäuser, 1997), 34–47, 70–73. The Club of Rome was officially set up in June 1970 in Geneva as a private nonprofit organi-zation according to Swiss law. The secretariat was initially based in Rome, then in Paris. Today the club is located in Winterthur, Switzerland.

9. William Watts, Foreword, in Meadows et al., *Limits*, 9–12.

10. Club of Rome, "Predicament of Mankind."

11. Club of Rome, "Predicament of Mankind," 13. See also Peccei, *Chasm*, xvi.

12. Club of Rome, "Predicament of Mankind," 9.

13. Aurelio Peccei, "Reflections on Bellagio," in *Perspectives of Planning: Proceedings of the OECD Working Symposium on Long-Range Forecasting and Planning, Bellagio*, ed. Erich Jantsch, 517–19 (Paris: OECD, 1969).

14. Erik P. Rau, "The Adoption of Operations Research in the United States during World War II," in *Systems, Experts, and Computers: The Systems Approach in Management and Engineering, World War II and after*, ed. Agatha C. Hughes and Thomas P. Hughes, 57–92 (Cambridge, MA: MIT Press, 2000), 77. See also Mike Fortun and S. S. Schweber, "Scientists and the Legacy of World War II: The Case of Operations Research (OR)," *Social Studies of Science* 23 (1993), 595–642.

15. Sonja M. Amadae, *Rationalizing Capitalist Democracy: The Cold War Origins of Rational Choice Liberalism* (Chicago: University of Chicago Press, 2003), 43.

16. The RAND Corporation, *The First Fifteen Years* (Santa Monica, 1963), 27; see Martin J. Collins, *Cold War Laboratory: RAND, the Air Force, and the American State, 1945–1950* (Washington: Smithsonian Institution Scholarly Press, 2002), 170–73. Michael Hagner, "Vom Aufstieg und Fall der Kybernetik als Universalwissenschaft," in *Die Transformation des Humanen: Beiträge zur Kulturgeschichte der Kybernetik*, ed. Michael Hagner and Erich Hörl, 38–72 (Frankfurt am Main: Suhrkamp, 2008). Seefried, *Zukünfte*, 59–62.

17. Peter Galison, "Computer Simulations and the Trading Zone," in *The Disunity of Science: Boundaries, Contexts, and Power*, ed. Peter Galison and David J. Stump, 118–57 (Stanford, CA: Stanford University Press, 1996). Paul N. Edwards, *The Closed World: Computers and the Politics of Discourse in Cold War America* (Cambridge, MA: MIT Press 1996), 120.

18. The Bellagio Declaration on Planning, in Jantsch, *Perspectives of Planning*, 7–9; Club of Rome, Statutes, June 16, 1970, in The Rockefeller Archive Center (RAC), Weiss Collection, 88/9. Also Club of Rome, "Predicament of Mankind." Cf. also Moll, *Scarcity*, 49–92.

19. Club of Rome, Peccei to Paul A. Weiss, February 16, 1970, in RAC, Weiss Collection, 89/1.

20. Club of Rome, "Predicament of Mankind." (1970) (compiled by Ozbekhan).

21. Cf. Peccei, *Human Quality*, 71–72. Friedemann Hahn, "Von Unsinn bis Untergang: Rezeption des Club of Rome und der Grenzen des Wachstums in der Bundesrepublik der frühen 1970er Jahre" (PhD diss., University of Freiburg, 2006), 49.

22. Edwards, *Closed World*, 75–110. For the transfer of military research findings to environmental research, see Jacob Darwin Hamblin, *Arming Mother Nature: The*

Birth of Catastrophic Environmentalism (Oxford: Oxford University Press, 2013) (on Forrester, see 173–78). Also Patrick Kupper, "Weltvernichtungsmaschinen: Die Bombe, die ökologische Revolution und die Transformation der Zukunft als Katastrophe," in *Die Krise der Zukunft II: Verantwortung und Freiheit angesichts apokalyptischer Szenarien,* ed. Jens Köhrsen, Harald Matern, and Georg Pfleiderer (Zurich: Nomos, forthcoming).

23. Jay W. Forrester, *Industrial Dynamics* (Cambridge, MA: MIT Press, 1966). Forrester, *Urban Dynamics* (Cambridge, MA: MIT Press, 1969). Moll, *Scarcity,* 72.

24. The Club of Rome, Peccei to Detlev Bronk, July 31, 1970, in RAC, Bronk Papers, 5/13.

25. Jay W. Forrester, *World Dynamics* (Cambridge, MA: Wright-Allen Press, 1971).

26. See Helga Nowotny, "Vergangene Zukunft: Ein Blick zurück auf die 'Grenzen des Wachstums,'" in *Impulse geben—Wissen stiften: 40 Jahre VolkswagenStiftung,* ed. VolkswagenStiftung, 655–94 (Göttingen: Vandenhoeck und Ruprecht, 2002).

27. Club of Rome, Aurelio Peccei to the Club members on February 9, 1972, in RAC, Weiss Collection, 89/1. For a detailed description of the development stage of the report, see Moll, *Scarcity,* 93–115.

28. *Die Grenzen des Wachstums, Bericht des Club of Rome zur Lage der Menschheit; Halte à la croissance? Enquête sur le Club de Rom et rapport sur le limites à la croissance; Los Límites del crecimiento: informe al Club de Roma sobre el predicamento de la humanidad; I limiti dello sviluppo: Rapporto del System Dynamics Group Massachusetts Institute of Technology (MIT) per il progetto del Club di Roma sui dilemmi dell'umanita; Seichō no genkairōma kurabu 'jinrui no kiki' repōto tankōbon.*

29. Meadows et al., *Limits,* 23. The report appeared a year later: Donella Meadows and Dennis Meadows, eds., *Toward Global Equilibrium: Collected Papers* (Cambridge, MA: MIT Press, 1973).

30. Meadows et al., *Limits,* 23.

31. Meadows et al., *Limits,* 24.

32. Meadows et al., *Limits,* 183–84.

33. Alexander King et al., "Commentary by the Club of Rome Executive Committee," in Meadows et al., *Limits,* 185–97, 190.

34. Meadows et al., *Limits,* 21.

35. Meadows et al., *Limits,* 25. Cf. 21, 88–89.

36. Meadows et al., *Limits,* 102–3, 121.

37. Meadows et al., *Limits,* 124–45.

38. Meadows et al., *Limits*, 155.

39. See also Blanchard, "Technoscientific," 101.

40. Meadows et al., *Limits*, 22.

41. Meadows et al., *Limits*, 93. On the history of the scenario technique, see Patrick Kupper, "Szenarien: Genese und Wirkung eines Verfahrens der Zukunftsbestimmung," in *Die Krise der Zukunft I: Apocalyptische Diskurse und ihre gesellschoftlichen Bedingunger und Wirkungen*, ed. Jen Köhrsen, Harald Matern, and Georg Pfleiderer (Zurich: Nomos, forthcoming).

42. Meadows et al., *Limits*, 88, 169.

43. Meadows et al., *Limits*, 125.

44. Meadows et al., *Limits*, 170, 175; on the "rhetoric of balance," see also Robert M. Collins, *More: The Politics of Economic Growth in Postwar America* (Oxford: Oxford University Press, 2000), 145–52.

45. *Neue Zürcher Zeitung* (*NZZ*), no. 362, August 6, 1972 (translated from German). The article was the beginning of a series of pieces both for and against the report. Further articles *NZZ*, no. 382, August 17, 1972; no. 405, August 31, 1972; no. 418, September 7, 1972; no. 449, September 26, 1972.

46. Among others, Claire Sterling, "The Fate of an Unplanned World," *Financial Times*, March 3, 1972; Robert Gillette, "The Limits to Growth: Hard Sell for a Computer View of Doomsday," *Science*, no. 175, March 10, 1972, 1088–92; "Another Whiff of Doomsday," *Nature*, no. 238, March 10, 1972; Thomas von Randow, "So geht die Welt zugrunde," *Die Zeit*, no. 11, March 17, 1972; "Die Grenzen des Wachstums: Eine warnende Studie des Massachusetts Institute of Technology," *NZZ*, no. 233, May 21, 1972.

47. Hugh S. D. Cole et al., eds., *Thinking about the Future: A Critique of The Limits to Growth* (London: Sussex University Press, 1973). Heinrich von Nussbaum, ed., *Die Zukunft des Wachstums: Kritische Antworten zum "Bericht des Club of Rome"*. (Düsseldorf: Bertelsmann, 1973). Andrew Weintraub et al., eds., *The Economic Growth Controversy* (White Plains, NY: International Arts and Sciences Press, 1973). Dennis L. Meadows and Horst-Eberhard Richter, eds., *Wachstum bis zur Katastrophe? Pro und Contra zum Weltmodell* (Stuttgart: Deutsche Verlags-Anstalt, 1974); Willem L. Oltmans, ed., *"Die Grenzen des Wachstums": Pro und Contra* (Reinbek bei Hamburg: Rowohlt, 1974). Aurelio Peccei and Manfred Siebker, *Die Grenzen des Wachstums: Fazit und Folgestudien* (Reinbek bei Hamburg: Rowohlt, 1974).

48. Streich, *30 Jahre Club of Rome*, 49. Higher and lower sales figures have also been circulated.

49. Anthony Tucker, "Predicament of Man," *The Guardian*, March 6, 1972.

50. Anthony Lewis, "Ecology and Politics I," *New York Times*, March 4, 1972, cited in Paul Sabin, *The Bet: Paul Ehrlich, Julian Simon, and our Gamble over Earth's Future* (New Haven, CT: Yale University Press, 2013), 82.

51. "Limits to Misconception," *The Economist*, March 11, 1972, 20–22, cited in Nicholas Georgescu-Roegen, *Energy and Economic Myths: Institutional and Analytical Economic Essays* (New York: Pergamon, 1976), 21. Carl Kaysen, "The Computer that Printed Out W*O*L*F*," *Foreign Affairs*, July 1972, 660–68. "Weltuntergangs-Vision aus dem Computer," *Der Spiegel*, no. 21, May 15, 1972. On the reaction of the American press, see W. Patrick McCray, *The Visioneers: How a Group of Elite Scientists Pursued Space Colonies, Nanotechnologies, and a Limitless Future* (Princeton, NJ: Princeton University Press, 2013), 34–37.

52. For example, by the 1970 American Nobel Memorial Prize in Economics winner, Paul A. Samuelson, in Oltmans, *Grenzen*, 40–43. Wilfried Beckermann from University College London also offers a detailed critique: Wilfried Beckermann, "Economists, Scientists, and Environmental Catastrophe," *Oxford Economic Papers* (New Series) 3, no. 24 (1972), 327–44. Cf. Sabin, *Bet*, 64, 88–95.

53. Thomas Robert Malthus, *An Essay on the Principle of Population* (London: Pickering, 1986 [1798]).

54. Samuelson, in Oltmans, *Grenzen*, 41–42 (translated from German).

55. Thomas Robertson, *The Malthusian Moment: Global Population Growth and the Birth of American Environmentalism* (New Brunswick, NJ: Rutgers University Press, 2012). Sabin, *Bet*.

56. Francis Sandbach, "The Rise and Fall of the Limits to Growth Debate," *Social Studies of Science* 8 (1978), 495–520, here 498–501.

57. Christopher Freeman, "Malthus with a Computer," *Futures* 6 (1973), 5–13.

58. Marie Jahoda, Introduction, *World Futures: The Great Debate*, ed. Christopher Freeman and Marie Jahoda, 1–6 (New York: Martin Robertson, 1978), 2. Cf. Cole et al., *Thinking*.

59. Johan Galtung, "'The Limits to Growth' and Class Politics," *Journal of Peace Research* 10 (1973), 101–14, here 104–5.

60. Hans Magnus Enzensberger, "A Critique of Political Ecology," *New Left Review* 1, no. 84 (March–April 1974), 3–31, here 10, 12 (English translation of "Zur Kritik der politischen Ökologie," *Kursbuch* 33 (1973), 1–42). Enzensberger borrowed the term "eco-industrial complex" from James Ridgeway (*The Politics of Ecology* [New York: Dutton, 1971]), who saw a continuity between this and the military-industrial complex of the 1950s and 1960s, and emphasized the role of the U.S. secretary of defense Robert McNamara.

61. Enzensberger, "Critique," 16.

62. Cf. Enzensberger, "Critique." Moll, *Scarcity*, 115–16. The Left in Italy held a similar opinion: Piccioni, "Forty Years," 29–38.

63. Amílcar Oscar Herrera et al., *Catastrophe or New Society? A Latin American World Model* (Ottawa: International Development Research Centre, 1976). The Argentine group had already criticized the first drafts of the Club of Rome in 1971: The Executive Committee of the Club of Rome: The new Threshold, February 1973, RAC, Bronk Papers, 5/10.

64. Meadows et al., *Limits*, 164.

65. Herrera et al., *Catastrophe*.

66. Meadows et al., *Limits*, 29.

67. Cf. Rüdiger Graf, *Öl und Souveränität: Petroknowledge und Energiepolitik in den USA und Westeuropa in den 1970er Jahren* (Munich: De Gruyter Oldenbourg, 2014). On "reduced oil production," see, for example, Hans Hüssy, president of the World Wildlife Fund, Switzerland, in *Panda* 1/1974, 1.

68. *Der Landbote*, no. 302, December 31, 1973 (translated from German).

69. Sabin, *Bet*, 96–100.

70. Ernst Fritz Schumacher, *Small Is Beautiful: Economics as if People Mattered* (New York: Blond and Briggs, 1973). Cf. Sabin, *Bet*, 101–3. Seefried, *Zukünfte*, 279–87.

71. Alexander King, *Another Kind of Growth: Industrial Society and the Quality of Life* (London: The David Davies Memorial Institute of International Studies, 1972).

72. Mihajlo D. Mesarović and Eduard Pestel, *Mankind at the Turning Point: The Second Report to the Club of Rome* (New York: Dutton, 1974), 1.

73. Cf. Leena Riska-Campbell, *Bridging East and West: The Establishment of the International Institute for Applied Systems Analysis (IIASA) in the United States Foreign Policy of Bridge Building, 1964–1972* (Helsinki: The Finnish Society of Science and Letters, 2011), 88–89, 199–201. Peccei to Gvishiani, March 6, 1973, Archives of Imperial College London, Gabor Collection, MC 14/2.

74. See Elke Seefried, "Der kurze Traum von einer Steuerung der Zukunft: Zukunftsforschung in West und Ost," in *Die Zukunft des 20. Jahrhunderts*, ed. Lucian Hölscher, 179–220 (Frankfurt am Main: Campus, 2017).

75. Donella Meadows et al., *Groping in the Dark: The First Decade of Global Modelling* (Chichester: Wiley, 1982). Eglė Rindzevičiūtė, *The Power of Systems: How Policy Sciences Opened Up the Cold War World* (Ithaca: Cornell University Press, 2016).

76. Among others, see Herbert Gruhl, *Ein Planet wird geplündert: Die Schreckensbilanz unserer Politik* (Frankfurt am Main: Fischer, 1975).

77. Silke Mende, *"Nicht rechts, nicht links, sondern vorn": Eine Geschichte der*

Gründungsgrünen (Munich: Oldenbourg, 2011), 292–99. With the publication of *Blueprint for Survival* the "Movement for Survival," arose, from which the "British People's Party," then the "Ecology Party" and later the "Green Party" originated; Christopher A. Rootes, "Britain: Greens in a Cold Climate," in *The Green Challenge: The Development of Green Parties in Europe*, ed. Dick Richardson and Chris Rootes, 66–90 (London: Routledge, 1995); Elke Seefried, "Towards the Limits to Growth? The Book and Its Reception in West Germany and Great Britain 1972/73," *Bulletin of the German Historical Institute London* vol. 33 (2011), no 1, 3–37, here 20–22. On the formation of a Green List in the French parliamentary elections of 1974, see Alistair Cole and Brian Doherty, "France: Pas comme les autres—the French Greens at the Crossroads," in *Bulletin of the German Historical Institute London* 1, no. 33 (2011), 45–64, here 48.

78. Lars-Göran Engfeldt, *From Stockholm to Johannesburg and Beyond* (Stockholm: Utrikesdepartementet, 2009). Thorsten Schulz-Walden, *Anfänge globaler Umweltpolitik: Umweltsicherheit in der internationalen Politik (1969–1975)* (Munich: Oldenbourg, 2013). Kai F. Hünemörder, *Die Frühgeschichte der globalen Umweltkrise und die Formierung der deutschen Umweltpolitik (1950–1973)* (Stuttgart: Franz Steiner, 2004).

79. Cf, Hünemörder, *Frühgeschichte*, 274–75. Seefried, *Zukünfte*, 456–58.

80. Postulat Letsch, "Gesamtenergiekonzeption," October 5, 1972, Official Bulletin of the Federal Assembly, National Assembly, 1972, 1800–1802.

81. Center of Research for the History and Sociology of Swiss Politics at the University of Bern, ed., *Schweizerische Politik im Jahre/Année politique suisse 1972* (Bern, 1972), 85.

82. Craig Murphy, *The Emergence of the NIEO Ideology* (Boulder, CO: Westview Press, 1984). John Toye and Richard Toye, *The UN and Global Political Economy: Trade, Finance, and Development* (Bloomington: Indiana University Press, 2004), 184–229. Gilbert Rist, *The History of Development: From Western Origins to Global Faith*, 4th ed. (London: Zed Books, 2014), 109–70.

83. Cf. Jan Tinbergen, in Oltmans, *Grenzen*, 28–32. Jan Tinbergen et al., eds., *RIO: Reshaping the International Order: A Report to the Club of Rome* (New York: Dutton, 1976). Moll, *Scarcity*, 184–94; Elke Seefried, "Globalized Science: The 1970s Futures Field," *Centaurus* 59 (2017), 40–57. Seefried, "Rethinking Progress: On the Origin of the Modern Sustainability Discourse, 1970–2000," *Journal of Modern European History* 13 (2015), 377–400. Iris Borowy, *Defining Sustainable Development for Our Common Future: A History of the World Commission on*

Environment and Development (Brundtland Commission) (London: Routledge, 2014).

84. Marc Frey, "Neo-Malthusianism and Development: Shifting Interpretations of a Contested Paradigm," *Journal of Global History* 6 (2011), 75–97, here 91. On the influence of *The Limits to Growth* on the conception of China's one-child policy, see Matthew Connelly, *Fatal Misconception: The Struggle to Control World Population* (Cambridge, MA: Harvard University Press, 2008), 340–41. Robertson, *Malthusian Moment*, 85–103, 126–200.

85. Joseph J. Corn, epilogue in *Imaging Tomorrow: History, Technology and the American Future*, ed. Joseph J. Corn (Cambridge, MA: MIT Press, 1986), 219.

86. For example Heinz-Ulrich Nennen, *Ökologie im Diskurs* (Opladen: Westdeutscher Verlag, 1991), 81–82.

87. On the warning prognosis, see Robert Merton, "The Self-Fulfilling Prophecy," *Antioch Review* 2, no. 8 (1948), 193–201, here 196.

88. Mesarović and Pestel, *Mankind*.

89. Adam Rome, "'Give Earth a Chance': The Environmental Movement and the Sixties," *Journal of American History* 90 (2003), 525–54, here 541–51. John McCormick, *Reclaiming Paradise: The Global Environmental Movement* (Bloomington: Indiana University Press, 1989), 46–68.

90. Collins, *More*. Matthias Schmelzer, *The Hegemony of Growth: The OECD and the Making of the Economic Growth Paradigm* (Cambridge: Cambridge University Press, 2016). On the theory that affluence opened up the space for "postmaterial thought," see Samuel P. Hays, "The Limits to Growth Issue," *Explorations in Environmental History: Essays*, ed. Samuel P. Hays and Joel A. Tarr, 3–23 (Pittsburgh: University of Pittsburgh Press, 1998), 9.

91. Rome, *Earth*, 527–29.

92. Rachel Carson, *Silent Spring* (New York: Houghton Mifflin, 1962). McCray, *Visioneers*, 22–23.

93. Richard N. L. Andrews, *A History of American Environmental Policy* (New Haven, CT: Yale University Press, 1999), 203–4. Hal Rothman, *The Greening of a Nation? Environmentalism in the U.S. since 1945* (Fort Worth, TX: Harcourt, 1997), 109–21.

94. Schulz-Walden, *Anfänge globaler Umweltpolitik*, 79–152, 177–243. Schmelzer, "Crisis before the Crisis." Kai F. Hünemörder, "Environmental Crisis and Soft Politics: Détente and the Global Environment, 1968–1975," in *Environmental Histories of the Cold War*, ed. John R. McNeill and Corinna R. Unger, 257–76 (New York: Cambridge University Press, 2010); Hünemörder, *Frühgeschichte*, 121–81.

95. *Nationalzeitung*, no. 489, October 24, 1971.

96. J. R. McNeill, "The Environment, Environmentalism, and International Society in the Long 1970s," in *The Shock of the Global: The 1970s in Perspective*, ed. Niall Ferguson et al., 262–78 (Cambridge, MA: Harvard University Press, 2010), 265.

97. Adam Rome, *The Genius of Earth Day: How a 1970 Teach-in Unexpectedly Made the First Green Generation* (New York: Hill and Wang, 2013). Thorsten Schulz, *Das Europäische Naturschutzjahr 1970: Versuch einer europaweiten Umweltkampagne*, working paper, WZB Berlin, 2006.

98. On the turn to environmental politics in the 1970s, see Patrick Kupper, "Die '1970er Diagnose': Grundsätzliche Überlegungen zu einem Wendepunkt der Umweltgeschichte," *Archiv für Sozialgeschichte* 43 (2003), 325–48. On Switzerland, see Patrick Kupper, *Atomenergie und gespaltene Gesellschaft: Die Geschichte des gescheiterten Projektes Kernkraftwerk Kaiseraugst* (Zurich: Chronos, 2003). On "ecological revolution" around the world about 1970, see Joachim Radkau, *Die Ära der Ökologie: Eine Weltgeschichte* (Munich: Beck, 2011), 134–68.

99. McNeill, "Environment," 264, 274–75. Hünemörder, *Frühgeschichte,* 166–69. Jens Ivo Engels, *Naturpolitik in der Bundesrepublik: Ideenwelt und politische Verhaltensstile in Naturschutz und Umweltbewegung 1950–1980* (Paderborn: Schöningh, 2006), 276–79.

100. Jakob Bächtold, "Wo steht der Naturschutz heute?" *Schweizer Naturschutz* 5 (1963), 123–25. Jakob Bächtold, "Naturschutz ist auch Menschenschutz," *Schweizer Naturschutz* 5 (1964), 121–24.

101. Kenneth E. Boulding, "The Economics of the Coming Spaceship Earth," in *Environmental Quality in a Growing Economy: Essays for the Sixth RFF Forum*, ed. Henry Jarrett, 3–14 (Baltimore: Johns Hopkins University Press, 1966). Barbara Ward, *Spaceship Earth* (New York: Columbia University Press, 1966).

102. Sabine Höhler, *Spaceship Earth in the Environmental Age, 1960–1990* (London: Pickering and Chatto, 2015). McCray, *Visioneers*, 23–24.

103. Alvin Toffler, *Future Shock* (New York: Random House, 1970). Paul R. Ehrlich, *The Population Bomb* (New York: Ballantine Books, 1968). Gordon Rattray Taylor, *The Doomsday Book: Can the World Survive?* (New York: World, 1970).

104. Barry Commoner, *Science and Survival* (New York: Viking Press, 1970). Cf. Michael Egan, *Barry Commoner and the Science of Survival: The Remaking of American Environmentalism* (Cambridge, MA: MIT Press, 2007).

105. Edward Goldsmith et al., "A Blueprint for Survival," preface in *The Ecologist* 2, no. 1 (1972), 1. Edward Goldsmith and Robert Allen, eds., *Blueprint for Survival* (Boston: Houghton Mifflin, 1972). The British government set up a "Committee on

Future World Trends," that commissioned its own world model in order to test the theses of *Blueprint for Survival* and *The Limits to Growth*: Her Majesty's Stationery Office, Cabinet Office, *Future World Trends: A Discussion Paper on World Trends in Population, Resources, Pollution etc., and Their Implications* (London: H.M. Stationery Office, 1976). Seefried, "Towards the Limits to Growth?" 34.

106. SPD Parteivorstand, "Sozialdemokratische Perspektiven," in *Perspektiven: Sozialdemokratische Politik im Übergang zu den siebziger Jahren*, ed. Horst Ehmke (Reinbek bei Hamburg: Rowohlt, 1969), 23 (translated from German). Cf. Glen O'Hara, *From Dreams to Disillusionment: Economic and Social Planning in 1960s Britain* (Basingstoke: Palgrave Macmillan, 2007). Gabriele Metzler, *Konzeptionen politischen Handelns von Adenauer bis Brandt: Politische Planung in der pluralistischen Gesellschaft* (Paderborn: Schöningh, 2005). Jon Agar, *The Government Machine: A Revolutionary History of the Computer* (Cambridge, MA: MIT Press, 2003).

107. Martel Gerteis, *Automation: Chancen und Folgen für Mensch, Wirtschaft und Politik* (Zurich: Forkel, 1964), 340–41 (translated from German). Very similarly, Karl Steinbuch, *Die informierte Gesellschaft: Geschichte und Technik der Nachrichtentechnik* (Stuttgart: Deutsche Verlags-Anstalt, 1966). On this subject, see also Alexander Schmidt-Gernig, "The Cybernetic Society: Western Future Studies of the 1960s and 1970s and their Predictions for the Year 2000," in *What the Future Holds: Insights from Social Science*, ed. Richard N. Cooper and Peter R. G. Layard, 233–59 (Cambridge, MA: MIT Press, 2002). Elke Seefried, "Steering the Future: The Emergence of Futures Research and Its Production of Expertise, 1950s to early 1970s," *European Journal of Futures Research* 2 (2013) (http://dx.doi.org/10.1007/s40309-013-0029-y).

108. Meadows et al., *Limits*, 122.

109. Ingo A. Schwarz and Rolf Kappel, *Systemforschung 1970–1980: Entwicklungen in der Bundesrepublik Deutschland: Materialien zu einem Förderungsschwerpunkt der Stiftung Volkswagenwerk* (Göttingen: Vandenhoeck und Ruprecht, 1981). In futures studies, *The Limits to Growth* became an important reference work for this movement toward a more rigorous, self-reflective methodology. Seefried, *Zukünfte*. Wendell Bell, "Futures Studies Comes of Age: Twenty-Five Years after *The Limits to Growth*," *Futures* 33 (2001), 63–76.

110. Mesarović and Pestel, *Mankind*. Dennis Gabor and Umberto Colombo with Alexander King and R. Galli, *Beyond the Age of Waste: A Report to the Club of Rome* (London: Pergamon, 1976). Ernst Ulrich von Weizsäcker et al., *Faktor vier: Doppelter Wohlstand—halbierter Naturverbrauch: Der neue Bericht an den Club*

of Rome (Munich: Droemer Knaur, 1995). Sabin, *Bet*, 131–80. Robertson, *Malthusian Moment*, 208–11.

111. Cf. Radkau, *Ära*, 580–613. Frank Uekötter, *Am Ende der Gewissheiten: Die ökologische Frage im 21. Jahrhundert* (Frankfurt am Main: Campus, 2011), 121–24. Mu Ramkumar, ed., *On a Sustainable Future of Earth's Natural Resources* (Berlin: Springer, 2013). For debate about the report thirty years later, see http://www.thesolutionsjournal.com/node/569; http://www.donellameadows.org/archives/a-synopsis-limits-to-growth-the-30-year-update/; and for forty years later, see http://cassandralegacy.blogspot.co.at/2011/06/limits-to-growth-revisited.html (accessed August 28, 2017). In addition, see the report by *The Limits to Growth* coauthor Jørgen Randers to the Club of Rome, Jørgen Randers, *2052: A Global Forecast for the Next Forty Years: A Report to the Club of Rome Commemorating the 40th Anniversary of* The Limits to Growth (White River Junction, VT: Chelsea Green, 2012).

112. Graham M. Turner, "A Comparison of *The Limits to Growth* with 30 Years of Reality," *Global Environmental Change* 18 (2008), 397–411. The author reaffirmed this assessment in an update in 2014: Graham Turner, *Is Global Collapse Imminent? An Updated Comparison of* The Limits to Growth *with Historical Data*, MSSI Research Papers (Research paper series), 2014.

113. Bjørn Lomborg and Olivier Rubin, "The Dustbin of History: The Limits to Growth," *Foreign Policy*, November 9, 2009 (http://foreignpolicy.com/2009/11/09/the-dustbin-of-history-limits-to-growth/; accessed August 28, 2017). See also Lomborg's article, "The Limits to Panic," June 17, 2013 (http://www.project-syndicate.org/commentary/economic-growth-and-its-critics-by-bj-rn-lomborg; accessed August 28, 2017).

114. Lomborg and Rubin, "Dustbin."

115. One of the simulated scenarios reckoned with unlimited natural resources. Meadows et al., *Limits*, 131–37.

116. Merton, "Self-Fulfilling Prophecy."

117. Randers, *2052*, 15–16.

4. THE SUM OF ALL GERMAN FEARS

1. Bartholomäus Grill, "Deutschland—ein Waldesmärchen," *Die Zeit*, no. 53, December 25, 1987. See also Erich Wiedemann, *Die deutschen Ängste: Ein Volk in Moll* (Frankfurt: Ullstein, 1990), 184.

2. Ursula Breymayer and Bernd Ulrich, eds., *Unter Bäumen: Die Deutschen und der*

Wald (Dresden: Sandstein, 2011). See also Johannes Zechner, *Der deutsche Wald: Eine Ideengeschichte 1800–1945* (Darmstadt: Philipp von Zabern, 2016).

3. Tobias Huff, *Natur und Industrie im Sozialismus: Eine Umweltgeschichte der DDR* (Göttingen: Vandenhoeck und Ruprecht, 2015), 309.

4. Cf. Laurent Schmit, "Le 'Waldsterben': Convergences et divergences franco-allemandes face à un problème écologique," in *Européanisation au XXe siècle: Un regard historique*, ed. Matthieu Osmont et al., 169–84 (Brussels: Peter Lang, 2012).

5. For a more comprehensive discussion, see Frank Uekötter, *Von der Rauchplage zur ökologischen Revolution: Eine Geschichte der Luftverschmutzung in Deutschland und den Vereinigten Staaten von Amerika 1880–1970* (Essen: Klartext, 2003), ch. 10.

6. Arne Andersen, *Historische Technikfolgenabschätzung am Beispiel des Metallhüttenwesens und der Chemieindustrie 1850–1933* (Stuttgart: Steiner, 1996), 57–60; Julius von Schroeder and Carl Reuss, *Die Beschädigung der Vegetation durch Rauch und die Oberharzer Hüttenrauchschäden* (Berlin: Parey, 1883; repr. Hildesheim: Olms, 1986).

7. *Stenographische Berichte über die Verhandlungen des Preußischen Hauses der Abgeordneten*, 19. Legislaturperiode, 3. Session 1901, vol. 2 (Berlin, 1901), ca. 1992.

8. Arnold Heller, "Die Planung neuer und die Erweiterung bestehender Industrieanlagen im Hinblick auf die Reinhaltung der Luft," *Gesundheits-Ingenieur* 71 (1950), 156–59, here 156. Similarly Ernst Effenberger, "Die Verunreinigung der Stadtluft durch industrielle Abgase und Geruchsstoffe," *Städtehygiene* 4 (1953), 321–29, here 326.

9. Robert E. Swain, "Smoke and Fume Investigations: A Historical Review," *Industrial and Engineering Chemistry* 41 (1949), 2384–88, here 2384. See also Moyer D. Thomas, "Auswirkungen der Luftverunreinigung auf Pflanzen," in *Die Verunreinigung der Luft*, ed. World Health Organization, 229–77 (Weinheim: Chemie, 1964), 229.

10. Wilhelm Liesegang, *Die Reinhaltung der Luft* (Leipzig, 1935), 11.

11. Sächsisches Hauptstaatsarchiv Dresden Finanzministerium no. 2891, 15–22.

12. Bundesarchiv (BArch) R 154/12132, p. 47.

13. BArch R 154/31, attachment to Rundschreiben der Wirtschaftsgruppe Metallindustrie, March 11, 1940, p. 7.

14. Geheimes Staatsarchiv Preußischer Kulturbesitz Berlin Rep. 120 BB II a 2 no. 13 adh., p. 11.

15. *Tonindustrie-Zeitung* 39 (1915), 430.

16. "die Erscheinungsform eines Gewohnheitsrechtes." BArch R 154/12026, Reichsanstalt für Wasser- und Luftgüte to Reichsminister des Innern, February 13, 1943, p. 9. See also H. B. Rüder, "Ausführung der Elektrofilter für Großkesselanlagen und die Sichtwirkung der Reingase," *Technische Mitteilungen* 34 (1941), 213–18, here 215.

17. Hauptstaatsarchiv Düsseldorf (HStAD) NW 354 No. 39, memorandum of November 17, 1959, p. 4.

18. E. O. Rasser, "Rauchschäden durch Rauchgifte und deren forstliche Behandlung," *Prometheus* 28 (1917), 90–92, here 91.

19. Karl Reuß, "Maßnahmen gegen die Ausbreitung von Hüttenrauchschäden im Walde," *VIII. Internationaler Landwirtschaftlicher Kongress Wien, Referate*, vol. 4, Sektion VIII Referat 5 (Vienna: Vernay, 1907), 2.

20. Schroeder and Reuss, *Beschädigung*, 11n.

21. Stadtarchiv Leipzig Stadtgesundheitsamt, no. 231, 60R.

22. W. Riede, "Ein einwandfreier Rauchschadennachweis," *Mitteilungen der Deutschen Landwirtschafts-Gesellschaft* 38 (1923), 423–24, here 423.

23. Paul Sorauer and Paul Graebner, *Handbuch der Pflanzenkrankheiten*, vol. 1: *Die nichtparasitären Krankheiten*, 5th ed. (Berlin: Parey, 1924), 836.

24. Schroeder and Reuss, *Beschädigung*, 12.

25. Hans Wislicenus, "Rauchschäden (Abgasfrage)," in *Industrielle Chemie*, ed. R. Escales, 163–96 (Stuttgart: Enke, 1912), 192.

26. Claus Ungewitter, *Verwertung des Wertlosen* (Berlin: Limpert, 1938), 243.

27. Rudolf Lorenz, "Rauchschäden," *Gesundheits-Ingenieur* 56 (1933), 449–53, here 451.

28. Gustav Lang, *Der Schornsteinbau. Viertes Heft: Sockel, Grundbau, Fuchs und Einstiegöffnungen, Bekämpfung der Rauch- und Russplage* (Hanover: Helwing, 1911), 550.

29. HStAD NW 85 no. 164, 118.

30. Uekoetter, *Age of Smoke*, 132–36.

31. *Wasser, Luft und Betrieb* 2 (1958), 180.

32. *Gesundheits-Ingenieur* 84 (1963), 28.

33. Gerhard Feldhaus and Horst D. Hansel, *Umweltschutz: Luftreinhaltung, Lärmbekämpfung. Rechts- und Verwaltungsvorschriften des Bundes und der Länder mit einer systematischen Einführung* (Cologne: Deutscher Gemeindeverlag, 1971), 40n.

34. BArch B 106/38378, Der Minister für Wirtschaft, Mittelstand und Verkehr des

Landes Nordrhein-Westfalen to Bundesminister für Gesundheitswesen, January 21, 1964, p. 4.

35. Bayerisches Hauptstaatsarchiv München MWi 28366, Luftreinhaltung und Lärmbekämpfung als Bestandteil der Struktur- und Gesundheitspolitik. Ausführungen des Arbeits- und Sozialministers Werner Figgen auf der Landespressekonferenz on May 7, 1968, p. 4.

36. K. Husmann and G. Hänig, "Das Grillo-AGS-Verfahren zur Entschwefelung von Abgasen," *Brennstoff—Wärme—Kraft* 23 (1971), 85–91, here 85. See also Werner Brocke, "Stand der Technik der Rauchgas- und Brennstoffentschwefelung—Pro und Kontra," *Luftverunreinigung* (1972), 13–22.

37. Hans Willi Thoenes, "Energieerzeugung und Immissionsschutz," *Technische Überwachung* 15 (1974), 415–20, here 417.

38. Norbert Haug, "Entschwefelung bei Kohlekraftwerken," *Umweltmagazin* 11, no. 7 (October 1982), 40–44, here 41; Ekkehard Richter, "Der technische Stand der Rauchgasentschwefelung in der Bundesrepublik Deutschland," *Umwelt* 14 (1984), 191–92, 195–96, 201–2, 205, here 192.

39. Jacob Jobst, "Regenschirme für unsere Bäume?" *Umweltmagazin* 11, no. 6 (September 1982), 17–20, here 20.

40. HStAD NW 354 no. 45, Niederschrift zur Sitzung des Hauptausschusses "Wirkung von Staub und Gasen," July 7, 1961, p. 4.

41. *Vorwärts*, no. 12 (March 21, 1974), 5.

42. Einheit. *Zeitschrift für Mitglieder der IG Bergbau und Energie* no. 20 (October 15, 1977), 1, 8.

43. Jobst, "Regenschirme," 20; *Der Spiegel*, no. 49 (November 30, 1981), 182.

44. HStAD NW 354 no. 42, Waldbauernverband Nordrhein-Westfalen to Landwirtschaftskammer Westfalen-Lippe, March 15, 1954.

45. Niedersächsisches Hauptstaatsarchiv Hannover Nds. 600 Acc. 153/92 no. 315, Deutscher Bundestag, Ausschuß für Gesundheitswesen, Ausschußdrucksache 52, August 14, 1958, p. 6.

46. BArch B 136/5344, Der Bundesminister für Ernährung, Landwirtschaft und Forsten to Bundesminister des Innern, March 29, 1971, p. 2.

47. See chapter 1 in this volume.

48. Rainer Grießhammer, *Letzte Chance für den Wald? Die abwendbaren Folgen des Sauren Regens* (Freiburg: Dreisam-Verlag, 1983), 80.

49. "Riesengeschütze für die Fernbeschießung größerer Waldgebiete." Hans Wislicenus, *Grundsätzliches zur technischen Abgas- und Rauchschädenfrage und zu deren Aussichten auf ihre Lösung* (Berlin: Chemie, 1933), 6.

50. W. Strewe, "Wärmeerzeugung mit festen Brennstoffen," *Gesundheits-Ingenieur* 86 (1965), 111–16, here 115.

51. Bayerisches Hauptstaatsarchiv München MArb 2596/I, Bayerisches Staatsministerium für Arbeit und soziale Fürsorge to Mineralölwirtschaftsverband, February 25, 1969, p. 2.

52. Wilhelm Knabe, "Luftverunreinigungen und Waldwirtschaft. Konsequenzen für die forstliche Forschung," *Berichte über Landwirtschaft* 50 (1972), 169–81, here 174.

53. S. Gericke and B. Kurmies, "Pflanzennährstoffe in den atmosphärischen Niederschlägen," *Die Phosphorsäure* 17 (1957), 279–300, here 289n.

54. F. C. Olds, "Air Pollution Control: Good Intentions in Search of Directions," *Power Engineering* 73, no. 9 (September 1969), 28–35, here 33. See also Karl-Friedrich Wentzel, *Was bleibt vom Waldsterben? Bilanz und Denkanstöße zur Neubewertung der derzeitigen Reaktion der Wälder auf Luftschadstoffe* (Freiburg: Hochschul-Verlag, 2001), 14n.

55. Rat von Sachverständigen für Umweltfragen, *Waldschäden und Luftverunreinigungen, Sondergutachten März 1983* (Stuttgart: Kohlhammer, 1983), 8.

56. Christer Agren, "Die Schäden durch Saure Niederschläge in Schweden," in *Das Waldsterben: Ursachen, Folgen, Gegenmaßnahmen*, 2nd ed., ed. Arbeitskreis Chemische Industrie, Katalyse-Umweltgruppe Köln, 98–99 (Cologne: Kölner Volksblatt-Verlag, 1984), 99.

57. Edda Müller, *Innenwelt der Umweltpolitik. Sozial-liberale Umweltpolitik—(Ohn)macht durch Organisation?* (Opladen: Westdeutscher Verlag, 1986), 224.

58. BArch B 295/7219, memorandum of Abteilungsleiter U, October 4, 1982, p. 1.

59. P. Davids, N. Haug, M. Lange, H. J. Oels, and B. Schmidt, "Luftreinhaltung bei Kraftwerks- und Industriefeuerungen," *Brennstoff—Wärme—Kraft* 33 (1981), 170–75, here 170; Jobst, "Regenschirme," 20.

60. BArch B 106/57482.

61. Birgit Metzger, "Rettet den Wald! Eine westdeutsche Debatte über ein Umweltproblem," in *Das Waldsterben: Rückblick auf einen Ausnahmezustand*, ed. Roderich von Detten, 64–81 (Munich: Oekom, 2013), 67.

62. Deutscher Bundestag, 7. Wahlperiode, *Drucksache 2802*, p. 177. See also Renate Mayntz, *Vollzugsprobleme der Umweltpolitik: Empirische Untersuchung der Implementation von Gesetzen im Bereich der Luftreinhaltung und des Gewässerschutzes* (Stuttgart: Kohlhammer, 1978).

63. Grießhammer, *Letzte Chance*, 71n.

64. Helmut Weidner and Peter Knoepfel, "Implementationschancen der EG-Richtlinie

zur SO$_2$-Luftreinhaltepolitik: Ein kritischer Beitrag zur Internationalisierung von Umweltpolitik," *Zeitschrift für Umweltpolitik* 4 (1981), 27–67, here 42.

65. Schroeder and Reuss, *Beschädigung*, 254. See also Uekötter, *Rauchplage*, ch. 10.2.

66. Martin Bemmann, "Das Waldsterben—ein 'modernes' Umweltproblem?" in von Detten, *Das Waldsterben*, 16–29, here 17. Cf. Roland Schäfer, *"Lamettasyndrom" und "Säuresteppe": Das Waldsterben und die Forstwissenschaften 1979–2007* (Freiburg: Institut für Forstökonomie, 2012), 63n.

67. Roland Wagner, "Vom Alarm zum etablierten Forschungsobjekt: Waldsterben in den Forstwissenschaften," in von Detten, *Das Waldsterben*, 34–47, here 34.

68. HStAD NW 610 AL 403, Landtag Nordrhein-Westfalen, 9. Wahlperiode, Ausschußprotokoll 9/652, Ausschuß für Ernährung, Land-, Forst- und Wasserwirtschaft, 50. Sitzung, May 27, 1982, p. 42. For similar remarks, see Schäfer, *"Lamettasyndrom,"* 199–207.

69. Silke Mende, *"Nicht rechts, nicht links, sondern vorn": Eine Geschichte der Gründungsgrünen* (Munich: Oldenbourg, 2011).

70. E. Gene Frankland, "Germany: The Rise, Fall and Recovery of Die Grünen," in *The Green Challenge: The Development of Green Parties in Europe*, ed. Dick Richardson and Chris Rootes, 23–44 (London: Routledge, 1995), 23n.

71. BArch B 342/916, letter of the Bund Natur- und Umweltschutz Nordrhein-Westfalen of March 29, 1982.

72. Hans Bibelriether, "Schutzwald—wogegen oder wofür?" in *Rettet den Wald*, ed. Horst Stern et al., 339–48 (Munich: Kindler, 1979), 345.

73. Egmont R. Koch and Fritz Vahrenholt, *Seveso ist überall: Die tödlichen Risiken der Chemie* (Cologne: Kiepenheuer und Witsch, 1978). On the Seveso affair, see Frank Uekötter and Claas Kirchhelle, "Wie Seveso nach Deutschland kam: Umweltskandale und ökologische Debatte von 1976 bis 1986," *Archiv für Sozialgeschichte* 52 (2012), 317–34.

74. *Der Spiegel*, no. 47 (November 16, 1981), 96, 101; no. 48 (November 23, 1981), 188, 198; no. 49 (November 30, 1981), 174. On the Cold War context of environmentalism, see Jacob Darwin Hamblin, *Arming Mother Nature: The Birth of Catastrophic Environmentalism* (Oxford: Oxford University Press, 2013).

75. Similarly, Hans-Jochen Luhmann, "Warum hat nicht der Sachverständigenrat für Umweltfragen, sondern der SPIEGEL das Waldsterben entdeckt?" in *Jahrbuch Ökologie 1992*, ed. Günter Altner et al., 292–307 (Munich: Beck, 1992), 295.

76. Rudi Holzberger, *Das sogenannte Waldsterben: Zur Karriere eines Klischees: Das Thema Wald im journalistischen Diskurs* (Bergatreute: Eppe, 1995), 70–76. See also Schäfer, *"Lamettasyndrom,"* 39.

77. Holzberger, *Das sogenannte Waldsterben*, 68.

78. P. Rippert, "Beiträge zur Beurteilung von Rauchschäden im rheinisch-westfälischen Industriegebiet," *Glückauf* 48 (1912), 1992–2000, 2026–37, here 2027.

79. Manfred Görtemaker, *Geschichte der Bundesrepublik Deutschland: Von der Gründung bis zur Gegenwart* (Munich: Beck, 1999), 592–96, 704–6; Hans-Peter Schwarz, *Helmut Kohl: Eine politische Biographie* (Munich: Deutsche Verlags-Anstalt, 2012), 326–41.

80. For a discussion of West German history as a permanent quest for safety and security, see Eckart Conze, *Die Suche nach Sicherheit: Eine Geschichte der Bundesrepublik Deutschland von 1949 bis in die Gegenwart* (Munich: Siedler, 2009).

81. *Der Spiegel*, no. 2 (1983), 32.

82. Jens Ivo Engels, "'Inkorporierung' und 'Normalisierung' einer Protestbewegung am Beispiel der westdeutschen Umweltproteste in den 1980er Jahren," *Mitteilungsblatt des Instituts für soziale Bewegungen* 40 (2008), 81–100, here 85.

83. *Süddeutsche Zeitung*, no. 25 (February 1, 1983), 24.

84. Deutscher Bundestag, 10. Wahlperiode, Drucksache no. 67, p. 2.

85. Hoimar von Ditfurth, *So laßt uns denn ein Apfelbäumchen pflanzen: Es ist soweit* (Hamburg: Rasch und Röhring, 1985), 120. Carl Amery, "Das Zeichen an der Wand," in *Das Waldsterben: Ursachen, Folgen, Gegenmaßnahmen*, 2nd ed., ed. Arbeitskreis Chemische Industrie, Katalyse-Umweltgruppe Köln, 11–13 (Cologne: Kölner Volksblatt-Verlag, 1984), 13.

86. *Nürnberger Zeitung*, no. 283 (December 7, 1983), 3.

87. Dietrich Thränhardt, *Geschichte der Bundesrepublik Deutschland 1949–1990* (Frankfurt: Suhrkamp, 1996), 323.

88. BArch B342/929, Bund für Umwelt und Naturschutz Deutschland to Firma Wald-Numen, November 6, 1987.

89. Cf. Joachim Radkau and Lothar Hahn, *Aufstieg und Fall der deutschen Atomwirtschaft* (Munich: Oekom, 2013).

90. For the broader context of German environmentalism, see Frank Uekötter, *The Greenest Nation? A New History of German Environmentalism* (Boston: MIT Press, 2014).

91. BArch B 342/929, pamphlet for Großkundgebung und Demonstrationszug gegen das Waldsterben, November 10, 1984.

92. BArch B 342/929, Deutsche Aktionsgemeinschaft Kampf gegen das Waldsterben, Pressemitteilung no. 1 of October 5, 1984, p. 2.

93. BArch B 106/72144, Bund für Umwelt und Naturschutz Deutschland, Presseinformation no. 152 (November 25, 1981).

94. Staatsarchiv Nürnberg Forstamt Feucht, no. 358, Bund Naturschutz in Bayern, Mit dem Wald stirbt unsere Zukunft, July 1983.

95. Birgit Metzger, *"Erst stirbt der Wald, dann Du!" Das Waldsterben als deutsches Politikum (1978–1986)* (Frankfurt: Campus, 2015), 337.

96. Jobst, "Regenschirme," 17.

97. BArch B 295/187, Referat U III 2, memorandum of June 30, 1983, p. 1. The official's nerves were strained as his administrative unit, which was in charge of forest death as well as maritime ecology, organic farming, and some other issues, was seriously understaffed, leaving him chronically overworked. (BArch B 295/187, Referat U II 1, memorandum of June 8, 1983.)

98. Arbeitskreis Chemische Industrie, *Das Waldsterben*, 77.

99. Stern et al., *Rettet den Wald*.

100. BArch B 295/187, Fraunhofer-Institut für Toxikologie und Aerosolforschung, Überblick über das Phänomen Waldsterben und die möglichen Ursachen, July 1983, p. 13. However, the authors pointed out that this would usually bring "a decrease of profits."

101. BArch B 102/285339, Arbeitsgemeinschaft deutscher Waldbesitzerverbände an alle Waldbesitzer, Bonn, September 1983, p. 7.

102. BArch B 102/285339, Arbeitsgemeinschaft deutscher Waldbesitzerverbände an alle Waldbesitzer, Bonn, September 1983, p. 9.

103. Cf. Walter Leisner, *Waldsterben: Öffentlich-rechtliche Ersatzansprüche* (Cologne: Heymann, 1983).

104. BArch B 102/285339, Deutscher Bauernverband to Bundeskanzler der Bundesrepublik Deutschland, December 21, 1983, p. 3.

105. BArch B 136/26017, Rede Bundeskanzler Dr. Helmut Kohl auf dem Waldbauerntag der Arbeitsgemeinschaft Deutscher Waldbesitzerverbände in Cologne on November 20, 1984, p. 8; Der Bundesminister des Innern to Philipp Freiherr von Boeselager, chairman of the Arbeitsgemeinschaft Deutscher Waldbesitzerverbände, December 17, 1984. For the stance of the forest owners, see BArch B 136/26017, Arbeitsgemeinschaft Deutscher Waldbesitzerverbände to Bundeskanzler Dr. Helmut Kohl, August 22, 1984, and Philipp Freiherr von Boeselager, "Gegen das Waldsterben ist noch viel mehr zu tun," *Frankfurter Allgemeine*, no. 169 (August 2, 1984), 6.

106. BArch B 136/26017, Rede Bundeskanzler Dr. Helmut Kohl auf dem Waldbauerntag der Arbeitsgemeinschaft Deutscher Waldbesitzerverbände in Cologne on November 20, 1984, p. 8.

107. BArch B 136/26014, Gesamtverband des Deutschen Steinkohlenbergbaus to

Staatssekretär Manfred Lahnstein, Chef des Bundeskanzleramts, March 24, 1982, p. 2.

108. BArch B 102/285339, Vereinigung Deutscher Sägewerksverbände to Bundesministerium für Wirtschaft, October 10, 1983.

109. *Umweltmagazin* 14, no. 5 (September 1985), p. 76.

110. Schäfer, *"Lamettasyndrom,"* 218.

111. Lothar A. König, "Gibt es einen Zusammenhang zwischen Umweltradioaktivität und Waldschäden?" *GIT Fachzeitschrift für das Laboratorium* 29 (1985), 1123–24, 1127–28, 1130–31, 1256–58, 1260–62, 1264–65.

112. Wagner, "Vom Alarm," 39; Schäfer, *"Lamettasyndrom,"* 129n.

113. BArch B 295/7219, memorandum of Abteilungsleiter U, May 29, 1982.

114. BArch B 295/7219, Referat U I 7, memorandum of November 25, 1982, p. 3. The prospective sulfur dioxide concentration at Buschhaus was 12,650 mg/m³ (*Der Spiegel*. no. 38 [September 20, 1982], 106, 108).

115. BArch B 116/46595, Bundesverband Bürgerinitiativen Umweltschutz, Stellungnahme zum Vorentwurf einer Verordnung zur Durchführung des Bundes-Immissionsschutzgesetzes (Verordnung über Großfeuerungsanlagen), Stand: December 1980, Bearbeiter: Thomas Schwilling, 4.

116. BArch B 116/46595, Bundesverband Bürgerinitiativen Umweltschutz, Stellungnahme des BBU zum Entwurf einer Großfeuerungsanlagen-Verordnung (Stand: September 24, 1982), November 1982 (quotation), 8.

117. *Umweltmagazin* 12, no. 5 (August 1983), 30; Metzger, *Erst stirbt*, 356.

118. Bernd Schärer and Harald Keiter, "Abgasreinigung bei Großfeuerungsanlagen," *Umweltmagazin* 13, no. 6 (October 1984), 14–16, 18, 21, here 15.

119. Bernd Schärer and Norbert Haug, "Teurer Strom durch Umweltschutz?" *Umweltmagazin* 18, no. 8 (August 1989), 32–34, here 33.

120. Schärer and Keiter, "Abgasreinigung," 16.

121. Umweltbundesamt, *Daten zur Umwelt. Der Zustand der Umwelt in Deutschland*, edition 1997 (Berlin, 1997), 135n.

122. Harald Menig, *Abgas-Entschwefelung und-Entstickung* (Wiesbaden: Deutscher Fachschriften-Verlag, 1987), 46.

123. BArch B 295/7219, Abteilungsleiter U to the minister of the interior via Staatssekretär Dr. Hartkopf, May 6, 1982, p. 2.

124. Cf. Uekötter, *Greenest Nation*, 109.

125. See, for instance, BArch B 116/46595, Stellungnahme des Bundesverbandes der Deutschen Industrie e.V. zum Referentenentwurf vom 24.09.1982 für eine Dreizehnte Verordnung zur Durchführung des Bundes-Immissionsschutzgesetzes

(Verordnung über Großfeuerungsanlagen—13. BImSchV), Cologne, November 19, 1982; B 116/46595, Vereinigung Industrielle Kraftwirtschaft to Bundesminister für Ernährung, Landwirtschaft und Forsten, May 16, 1983; B 295/7219, memorandums of August 12, 1982, and January 31, 1983.

126. BArch B 295/7219, Abteilungsleiter U, memorandum of May 5, 1983.

127. Cf. Hendrik Ehrhardt, "Umweltpolitik in der Stromwirtschaft: Vom Kostentreiber zur Legitimationsinstanz?" *Zeitschrift für Unternehmensgeschichte* 60 (2015), 194–217.

128. Umweltbundesamt, *Daten zur Umwelt*, 135n. In order to allow comparisons, figures for 1990 and 1994 do not include the former GDR.

129. Menig, *Abgas-Entschwefelung*, 46.

130. *Umweltmagazin* 12, no. 8 (December 1983), 32; 17, no. 10 (October 1988), 65.

131. Bernd Schärer, "Abgaskatalysator—ein schwacher EG-Kompromiß," *Umweltmagazin* 14, no. 3 (May 1985), 24–26, here 25; Jacob Jobst, "Mehr Luft zum Leben," *Umweltmagazin* 13, no. 7 (November 1984), 22–23, here 23.

132. Schärer, "Abgaskatalysator."

133. Helmut Weidner, "17 Länder im Vergleich. Luftreinhaltepolitik in Europa— Leistungen und Möglichkeiten," *Umweltmagazin* 15, no. 7 (November 1986), 26–29; 15, no. 8 (December 1986), 36–39, here 28.

134. Schärer, "Abgaskatalysator," 26.

135. Fortbildungszentrum Gesundheits- und Umweltschutz Berlin, *Auswertung der Waldschadensforschungsergebnisse (1982–1992) zur Aufklärung komplexer Ursache-Wirkungsbeziehungen mit Hilfe systemanalytischer Methoden* (Berlin: Erich Schmidt, 1997), p. 9.

136. Cf. Wagner, "Vom Alarm," 40, 43.

137. BArch B 136/26017, Rede Bundeskanzler Dr. Helmut Kohl auf dem Waldbauerntag der Arbeitsgemeinschaft Deutscher Waldbesitzerverbände in Cologne on November 20, 1984, p. 7. For a comprehensive discussion of that notion, see Uekötter, *Greenest Nation*.

138. BArch B 106/70231, Der Bundesminister des Innern to die für den Immissionsschutz zuständigen obersten Landesbehörden, January 11, 1980, and Der Bundesminister des Innern to Umweltbundesamt, May 23, 1980.

139. BArch B 106/69328.

140. Department of Special Collections, Charles E. Young Research Library, University of California, Los Angeles, Collection 1199 Box 85 Folder 4, Coalition to Tax Pollution, The Sulfur Tax, May 25, 1972.

141. Huff, *Natur und Industrie*, 134–41.

142. BArch B 295/187, Bundesrat Drucksache 43/1/83 of July 4, 1983.

143. Peter Menke-Glückert, "Vorhaben der Bundesregierung zur Umweltschutz-Gesetzgebung im Hinblick auf die Energiewirtschaft," *VGB Kraftwerkstechnik* 61 (1981),540–42, here 542.

144. Erwin Nießlein, "Das Ausmaß der Walderkrankung," in *Was wir über das Waldsterben wissen*, ed. Erwin Nießlein and Gerhard Voss, 14–25 (Cologne: Deutscher Instituts-Verlag, 1985), 14.

145. H. Stratmann, "Begrüßung," in *Waldschäden*, ed. VDI-Kommission Reinhaltung der Luft, 1–3 (Düsseldorf: VDI-Verlag, 1985), 1.

146. Heinz Ellenberg, "Blatt und Nadelverlust oder standörtlich wechselnde Ausbildung des Photosynthese-Apparats? Fragen zum Waldschadenbericht 1992," *Schweizerische Zeitschrift für Forstwesen* 145 (1994), 413–16, here 413. See also Heinz Ellenberg, "Neuartige Waldschäden? Ökologische Kritik an den Untersuchungsmethoden," *Allgemeine Forst Zeitschrift*, no. 15 (1995), 1–2.

147. Schäfer, "*Lamettasyndrom,*" 343.

148. Rat von Sachverständigen für Umweltfragen, *Waldschäden und Luftverunreinigungen*, p. 11.

149. Peter Schütt et al., *So stirbt der Wald. Schadbilder und Krankheitsverlauf* (Munich: BLV-Verlagsgesellschaft, 1983), 89.

150. Erwin Nießlein, "Stand der Ursachenforschung," in Nießlein and Gerhard Voss, *Was wir*, 26–62, here 38.

151. Forschungsbeirat Waldschäden/Luftverunreinigungen der Bundesregierung und der Länder, *2. Bericht*, Karlsruhe, May 1986, p. 218.

152. Wagner, "Vom Alarm," 37. See also Schäfer, "*Lamettasyndrom,*" 180–82.

153. Martin Trampe-Oloff, "Zur Komplexität als Hindernis problemorientierter Reaktion auf das Waldsterben" (PhD diss., Freiburg University, 1985).

154. J. Evers, C. Franz, F. Cörver, and C. Ziegler, *Waldbäume: Bilderserien zur Einschätzung von Kronenverlichtung bei Waldbäumen* (Kassel: Faste, n.d.).

155. *Bericht über den Zustand des Waldes 1991* (Schriftenreihe des Bundesministers für Ernährung, Landwirtschaft und Forsten, Angewandte Wissenschaft no. 405, Münster-Hiltrup: Landwirtschaftsverlag, 1992), p. 23.

156. *Bericht über den Zustand des Waldes. Ergebnisse der Waldschadensforschung* (Schriftenreihe des Bundesministers für Ernährung, Landwirtschaft und Forsten, Angewandte Wissenschaft no 390, Münster-Hiltrup: Landwirtschaftsverlag, 1990).

157. Cf. Frank Uekötter, "Ökologische Verflechtungen. Umrisse einer grünen Zeitgeschichte," in *Geteilte Geschichte. Ost- und Westdeutschland 1970–2000*, ed. Frank Bösch, 117–52 (Göttingen: Vandenhoeck und Ruprecht, 2015), 124.

158. Siegfried Anders and Gerhard Hofmann, "Vom Nährstoff zum Schadstoff. Auswirkungen von Stickstoffeinträgen in mitteleuropäische Kiefernökosysteme," *Robin Wood Magazin*, no. 50 (1996), 33–35, here 34.

159. *Waldzustandsbericht. Ergebnisse der Waldschadenserhebung 1989* (Schriftenreihe des Bundesministers für Ernährung, Landwirtschaft und Forsten no. 381, Münster-Hiltrup: Landwirtschaftsverlag, 1990), 15.

160. Cf. Siegfried Anders, "Der Wandel in der Ökologie der Wälder—Herausforderung für eine zukunftsorientierte Waldforschung," Patient Wald—sterbenskrank oder kerngesund? (*Journalistenseminar der Information Umwelt, GSF-Forschungszentrum für Umwelt und Gesundheit* vol. 20, Neuherberg, 1996), 47.

161. Ellenberg, "Blatt und Nadelverlust," 415.

162. For an extensive discussion of this issue, see Roderich von Detten, "Umweltpolitik und Unsicherheit: Zum Zusammenspiel von Wissenschaft und Umweltpolitik in der Debatte um das Waldsterben der 1980er Jahre," *Archiv für Sozialgeschichte* 50 (2010), 217–69.

163. On the Sandoz fire, see Nils Freytag, "Der rote Rhein: Die Sandoz-Katastrophe vom 1. November 1986 und ihre Folgen," Themenportal Europäische Geschichte (2010), (http://www.europa.clio-online.de/site/lang__de/ItemID__459/mid__11 428/40208214/default.aspx; accessed February 28, 2015).

164. *Frankfurter Rundschau*, no. 153 (July 6, 1998), 22.

165. Hans-Joachim Fietkau, Heide Matschuk, Helmut Moser, and Wolfgang Schulz, *Waldsterben: Urteilsgewohnheiten und Kommunikationsprozesse—Ein Erfahrungsbericht* (Internationales Institut für Umwelt und Gesellschaft, UG-Report 86–6, Berlin, 1986), 8, 11.

166. BArch B 136/26017, Rede Bundeskanzler Dr. Helmut Kohl auf dem Waldbauerntag der Arbeitsgemeinschaft Deutscher Waldbesitzerverbände in Cologne on November 20, 1984, p. 2.

167. http://www.wahlen.bayern.de/vb-ve/index.php; accessed May 29, 2015.

168. https://www.statistik.bayern.de/presse/archiv/2009/vob_2_2009.php and https: //www.statistik.bayern.de/presse/archiv/2010/voe_4_2010.php; accessed May 30, 2015). One of the authors was among those who signed up and cast a vote; he did not live in Bavaria in 2004.

5. THE ENDANGERED AMAZON RAIN FOREST IN THE AGE OF ECOLOGICAL CRISIS

1. The term "tropical rainforest" was first used in 1898 by the German botanist Andreas F. W. Schimper. Andreas F. W. Schimper, *Plant Geography upon a Physiological Basis* (Oxford: Clarendon, 1903).

2. For contemporary discussions about the internationalization of the Amazon, see Alexei Barrionuevo, "Whose Rain Forest Is This, Anyway?" *New York Times*, May 18, 2008.

3. Global Witness Report, "Deadly Environment: The Rise in Killing of Environmental and Land Defenders" (https://www.globalwitness.org/en/campaigns/environmental-activists/deadly-environment/).

4. Cf. Jürgen Scheffran, *Climate Change, Human Security and Violent Conflict: Challenges for Societal Stability* (Berlin: Springer, 2012).

5. Alina Schadwinkel, "Indonesien erstickt am Smog," *Die Zeit*, November 4, 2015. Translations to English are the author's unless otherwise indicated.

6. George Monbiot, "Indonesia Is Burning: So Why Is the World Looking Away?" *The Guardian*, October 30, 2015.

7. Frank Uekötter, *Am Ende der Gewissheiten: Die ökologische Frage im 21. Jahrhundert* (Frankfurt am Main: Campus, 2011), 229.

8. Michael Tilly, "Kurze Geschichte der Apokalyptik," *Aus Politik und Zeitgeschichte* 62, no. 51–52 (2012), 18.

9. Susanna B. Hecht and Alexander Cockburn, *The Fate of the Forest: Developers, Destroyers, and Defenders of the Amazon* (Chicago: University of Chicago Press, 2010), 14–15.

10. Cf. Susanna B. Hecht, *The Scramble for the Amazon and the "Lost Paradise" of Euclides da Cunha* (Chicago: Chicago University Press, 2013).

11. Susanna B. Hecht, "The Last Unfinished Page of Genesis: Euclides da Cunha and the Amazon," *Novos Cadernos NAEA* 11, no. 1 (2008), 23.

12. Cf. Seth Garfield, *In Search of the Amazon: Brazil, the United States, and the Nature of a Region* (Durham, NC: Duke University Press, 2013).

13. Greg Grandin, *Fordlandia: The Rise and Fall of Henry Ford's Forgotten Jungle City* (New York: Metropolitan Books, 2009). Lúcio Flávio Pinto, *Jari: Toda a verdade sobre o projeto de Ludwig: As relações entre estado e multinacional na Amazônia* (São Paulo: Marco Zero, 1986).

14. One of these programs was led by the U.S. anthropologist Charles Wagley and commissioned by the Institute of Inter-American Affairs. Alfredo Wagner Berno

de Almeida, *Antropologia dos archivos da Amazônia* (Rio de Janeiro: Fundação Universidade do Amazonas, 2008), 163.

15. On the relevance of U.S. credit providers at the beginning of the military regime, see Thomas E. Skidmore, *Politics of Military Rule in Brazil 1964–85* (New York: Oxford University Press, 1988), 38–39.

16. Antoine Acker, *Volkswagen in the Amazon: The Tragedy of Global Development in Modern Brazil* (Cambridge: Cambridge University Press, 2017), 81.

17. Referring to Barry Commoner, Rachel Carson, and Aldo Leopold, M. Jimmie Killingsworth and Jacqueline S. Palmer have argued that "the politically motivated writing of a few well-respected and talented representatives of the scientific community" has been more important "than the effort of environmentalists to assimilate scientific findings." Killingsworth and Palmer, *Ecospeak: Rhetoric and Environmental Politics in America* (Carbondale: Southern Illinois University Press, 1992), 51.

18. Joachim Radkau, *The Age of Ecology: A Global History* (Malden, MA: Polity Press, 2014), 91.

19. Robin L. Chazdon, Introduction, in *Foundations of Tropical Forest Biology: Classic Papers with Commentaries*, ed. T. C. Whitmore and Robin L. Chazdon (Chicago: University of Chicago Press, 2002), 3.

20. María V. Secreto, "A ocupação dos 'espaços vazios' no governo Vargas: do 'Discurso do rio Amazonas à saga dos soldados da borracha," *Estudos Históricos* 40 (2007), 120.

21. The main work that led the way for the military regime's Amazon policy was Golbery do Couto e Silva, *Geopolítica do Brasil* (Rio de Janeiro: José Olympio, 1966). For concrete development plans, see Superintendência do desenvolvimento da Amazônia (SUDAM), ed., *Amazonia legal: Manual do Investidor* (Belém, 1972). SUDAM, ed., *Plano de desenvolvimento da Amazônia 1972–1974* (Belém, 1971).

22. Stefan Zweig, *Brasilien: ein Land der Zukunft* (Stockholm: Bermann-Fischer, 1941).

23. For one example, after his expedition to the Amazon Theodore Roosevelt stated: "Surely such a rich and fertile land cannot be permitted to remain idle, to lie as a tenantless wilderness, while there are such teeming swarms of human beings in the overcrowded, overpeopled countries of the Old World." Theodore Roosevelt, *Through Brazilian Wilderness* (Hamburg: Severus, 2013 [1914]), 298–99.

24. Emilio Moran, "The Transamazon Highway and Amazonian Development: Goals, Implementation, and the Reality 20 Years later," *Secolas Annals 20* (March 1989), 48.

25. Sue Branford and Oriel Glock, *The Last Frontier: Fighting over Land in the Amazon* (London: Zed Books, 1985).

26. Richard Bourne, *Assault on the Amazon* (London: Victor Gollancz, 1978), 34.

27. Bourne, *Assault on the Amazon*, 34.

28. Paul W. Richards, *The Tropical Rain Forest: An Ecological Study* (Cambridge: Cambridge University Press, 1952), 12.

29. Margaret E. Keck and Kathryn Sikkink have referred to later coalitions between western academics and local Amazonian activists as *advocacy networks*. I prefer the term *concerned scientists* with respect to environmental alarmism, given that many scholars were worried about tropical deforestation without necessarily establishing networks with Brazilian activists. Cf. Keck and Sikkink, eds., *Activists beyond Borders: Advocacy Networks in International Politics* (Ithaca, NY: Cornell University Press, 1998).

30. Norman Myers, "National Parks in Savannah Africa," *Science* 178, no. 4067 (1972), 1262n1.

31. Norman Myers, "The Global Problem of Tropical Deforestation," in *Proceedings of U.S. Strategy Conference on Tropical Deforestation* (Washington, DC: Department of State, 1978), 19–22.

32. Norman Myers, "The Hamburger Connection: How Central America's Forests Become North America's Hamburgers," *Ambio* 10, no. 1 (1981), 3–8.

33. Norman Myers, "Depletion of Tropical Moist Forests: A Comparative Review of Rates and Causes in the Three Main Regions," *Acta Amazonica* 12, no. 4 (1983), 745–58.

34. Norman Myers, *The Primary Source: Tropical Forests and Our Future* (New York: Norton, 1984).

35. Beginning at the end of the 1970s, Myers carried out numerous studies on behalf of the International Union for Conservation of Nature and advised institutions such as the U.S. National Academy of Sciences, the World Resources Institute, the World Bank, various UN Agencies, and the Intergovernmental Panel on Climate Change (https://fds.duke.edu/db/Nicholas/esp/faculty/normyers/files/CV.pdf).

36. Cyrus B. Dawsey III, "Geography," in *Envisioning Brazil: A Guide to Brazilian Studies in the United States, 1945–2003*, ed. Marshall C. Eakin and Paulo Roberto de Almeida (Madison: University of Wisconsin Press, 2005), 330.

37. The first such instance concerned the systematic enslavement of indigenous workers by the Peruvian Rubber Company in Peru's Putumayo region, as these atrocities were revealed and documented between 1909 and 1913 by the Amer-

ican engineer W. E. Hardenburg and the British consul Roger Casement. The second case was related to the genocidal crimes against indigenous people with the proven involvement of the Brazilian Indian Protection Service in the late 1960s. Cf. Michael T. Taussig, "Culture of Terror—Space of Death—Casement, Roger: Putumayo Report and the Explanation of Torture," *Comparative Studies in Society and History* 26, no. 3 (1984), 467–97. On the Figueiredo Report, see Jonathan Watts and Jan Rocha, "Brazil's 'Lost Report' into Genocide Surfaces after 40 Years," *The Guardian*, May 29, 2013.

38. Cf. Charles H. Wood and Marianne Schmink, *Contested Frontiers in Amazonia* (New York: Columbia University Press, 1992). Hecht and Cockburn, *Fate of the Forest.*

39. Radkau, *Age of Ecology*, 172.

40. In fact, the problematization of deforestation in Brazil reaches further back in history than one might expect. Cf. José A. Pádua, *Um Sorpo de Destruição: Pensamento político e crítica ambiental no Brasil escravista (1786–1888)* (Rio de Janeiro: Jorge Zahar, 2004).

41. Cf. Arturo Gómez-Pompa, Carlos Vázquez Yanes, and Sergio Guevara, "The Tropical Rain Forest: A Non-renewable Resource," *Science* 177, no. 4051 (1972), 762–65. Charles Wagley, *Man in the Amazon* (Gainesville: University of Florida Press, 1974).

42. Several articles on the Transamazônica were published in the German weekly magazine *Der Spiegel* and the U.S. magazine *Time*, for instance.

43. Bourne, *Assault on the Amazon*, 283.

44. Raimundo Pereira Rodrigues and Sérgio Buarque, "A colonização da Amazonia: Calor e fotosíntese, lendas e mistérios," *Opinião* 44 (1973), 4. Raimundo Pereira Rodrigues and Sérgio Buarque, "A teoria do cipó," *Opinião* 45 (1973), 3.

45. Branford and Glock, *Last Frontier*, 75.

46. Branford and Glock, *Last Frontier*, 76.

47. For example, at the beginning of the 1970s the German geographer Wolfgang Brücher and his Dutch colleague Jan Kleinpenning were among those who spoke out in favor of cattle-raising in the Amazon. Acker, *Volkswagen in the Amazon*, 62–63.

48. Betty J. Meggers, *Amazonia: Man and Culture in a Counterfeit Paradise* (Washington, DC: Smithsonian Institution Press, 1971), viii.

49. Some scholars criticized Meggers for assumptions that drew on the idea of the pristine, untouched Amazon rain forest. Cf. Candace Slater, *Entangled Edens: Visions of the Amazon* (Berkeley: University of California Press, 2002), 265.

50. Meggers, *Amazonia*, 155.

51. William M. Denevan, "Development and the Imminent Demise of the Amazon Rain Forest," *Professional Geographer* 25, no. 2 (1973), 130–35.

52. Cf. Philip M. Fearnside, "The Main Resources of Amazonia," paper presented at the Latin American Studies Association (LASA), Twentieth International Congress, Guadalajara, Mexico, April, 17–19, 1997 (http://www.biblioteca.clacso.edu. ar/ar/libros/lasa97/fearnside.pdf), 5.

53. Gómez-Pompa, Yanes, and Guevara, "Tropical Rain Forest," 762–65, 762.

54. Radkau, *Age of Ecology*, 172.

55. Cf. J. M. G. Kleinpenning, *The Integration and Colonization of the Brazilian Portion of the Amazon Basin* (Nijmegen: Institute of Geography and Planning, 1975), 150. Robert J. Goodland and Howard S. Irwin, *Amazon Jungle: Green Hell to Red Desert? An Ecological Discussion of the Environmental Impact of the Highway Construction Program in the Amazon Basin* (New York: Elsevier Scientific, 1975), 101.

56. Goodland and Irwin, *Amazon Jungle*, 1.

57. Hecht, "Last Unfinished Page," 9–10.

58. Cf. David Takacs, *The Idea of Biodiversity: Philosophies of Paradise* (Baltimore: Johns Hopkins University Press, 1996).

59. Philip M. Fearnside, "Deforestation and International Economic Development Projects in Brazilian Amazonia," *Conservation Biology* 1, no. 3 (October 1987), 220.

60. Cf. Harald Sioli and Gerd Kohlhepp, *Gelebtes, geliebtes Amazonien: Forschungsreisen im brasilianischen Regenwald zwischen 1940 und 1962* (Munich: Friedrich Pfeil, 2007).

61. F. Schaller, "Nachruf/Obituary Prof. Dr. Harald Felix Ludwig Sioli (1910–2004)," *Amazoniana* 18, no. 1/2 (December 2004), 163–68.

62. Elke Maier, "Abenteuer am Amazonas," *MaxPlanckForschung* 1 (2010), 93.

63. Maier, "Abenteuer am Amazonas," 93.

64. *New York Times* articles dealing with the environment rose from 150 to 1,700 per year between 1960 and 1970. Margaret R. Biswas and Asit K. Biswas, "Environment and Sustainable Development in the Third World: A Review of the Past Decade," *Third World Quarterly* 4 (1982), 481.

65. William H. McNeill, "Historical Perspective," in *Resources for an Uncertain Future: Papers Presented at a Forum Marking the Twenty-fifth Anniversary of Resources for the Future*, ed. Charles J. Hitch (Baltimore: Johns Hopkins University Press, 1978), 59.

66. Rhett Butler, Calculating Deforestation Figures for the Amazon (last updated January 26, 2017). http://rainforests.mongabay.com/amazon/deforestation_calculations.html.

67. German Bundestag, ed., *Protecting the Tropical Forests: A High Priority International Task. Second Report of the Enquete-Commission of the 11th German Bundestag "Preventive Measures to Protect the Earth's Atmosphere"* (Bonn: Deutscher Bundestag Referat Öffentlichkeitsarbeit, 1990), 114.

68. Cf. William Boyd, "Ways of Seeing in Environmental Law: How Deforestation Became an Object of Climate Governance," *Ecology Law Quarterly* 37, no. 3 (2010), 843–916.

69. Goodland and Irwin, *Amazon Jungle*, 36–37.

70. "As denúncias sobre a Amazônia que não foram publicadas," *Jornal da Tarde*, October 27, 1976.

71. Goodland and Irwin, *Amazon Jungle*, 2.

72. Goodland and Irwin, *Amazon Jungle*, i.

73. "Verbrannte Erde," *Der Spiegel* 12 (1973), 178.

74. Robert Allen, "The Year of the Rain Forest," *New Scientist* (April 24, 1975), 178; Edward Goldsmith, Robert Allen et al., "A Blueprint for Survival," *Ecologist* 2, no. 1 (January 1972).

75. As late as 2003, the U.S. geographer Michael Williams spoke of an "astonishing uncertainty and debate" regarding the rates, dimensions, and definitions of global deforestation. Williams, *Deforesting the Earth: From Prehistory to Global Crisis* (Chicago: University of Chicago Press, 2003), 446.

76. William C. Paddock, cited in "Durch die Wildnis," *Der Spiegel* 47 (1971), 145.

77. Wilhelm Brinkmann, cited in, "Verbrannte Erde," *Der Spiegel* 12 (1973), 178.

78. Irving Friedman, "The Amazon Basin, Another Sahel?" *Science* 197, no. 4298 (1977), 7.

79. Ernst J. Fittkau, "Tropical Rainforest," *Grzimek's Encyclopedia of Ecology*, ed. Bernhard Grzimek, Joachim Illies, and Wolfgang Klausewitz (London: Van Nostrand Reinhold, 1976), 313.

80. Ludwig Beck, "Ökosystem amazonischer Regenwald. Droht ein Kreislaufkollaps?" *Bild der Wissenschaft* 4 (1974), 48.

81. Karl Goldstein, "The Green Movement in Brazil," *The Green Movement Worldwide*, ed. Matthias Finger (Greenwich, CT: JAI Press, 1992), 134.

82. Norman Myers, "Tropical Forests, a Treasure House under Siege," *IUCN Parks* (March 1980), 4.

83. "NASA Tropical Deforestation Research," https://earthobservatory.nasa.gov/Features/Deforestation/deforestation_update4.php .

84. David Skole and Compton Tucker, "Tropical Deforestation and Habitat Fragmentation in the Amazon: Satellite Data from 1978 to 1988," *Science* 260, no. 5116 (1993), 1905–10.

85. Cf. Raoni Rajão, "Representations and Discourses: the Role of Local Accounts and Remote Sensing in the Formulation of Amazonia's Environmental Policy," *Environmental Science and Policy* 30 (2013), 60–71.

86. Wolfgang Sachs, "Satellitenblick: Die Ikone vom blauen Planeten und ihre Folgen für die Wissenschaft," *Technik ohne Grenzen*, ed. Ingo Braun and Bernward Joerges (Frankfurt am Main: Suhrkamp, 1994), 338–39.

87. Branford and Glock, *Last Frontier*, 321.

88. Andrew Revkin, *The Burning Season: The Murder of Chico Mendes and the Fight for the Amazon Rain Forest* (Washington, DC: Island Press, 2004), 231.

89. Food and Agriculture Organization, Forestry Department, "Comparison of Forest Area and Forest Area Change Estimates Derived from FRA 1990 and FRA 2000," Working Paper 59 (2001), 30 (ftp://ftp.fao.org/docrep/fao/006/ad068e/AD068E.pdf; accessed on April 20, 2015). German Bundestag, *Protecting the Tropical Forests*, 113.

90. Adrian Sommer, "Attempt at an Assessment of the World's Tropical Moist Forests," *Unasylva* 112–13 (1976), 5–27. Frédéric Achard et al., "Determination of Tropical Deforestation Rates and Related Carbon Losses from 1990 to 2010," *Global Change Biology* 20, no. 8 (2014), 2540–54.

91. Birgit Metzger, *"Erst stirbt der Wald, dann Du!" Das Waldsterben als deutsches Politikum (1978-1986)* (Frankfurt: Campus, 2015), 432.

92. See Kenneth Anders and Frank Uekötter, chapter 4 in this volume.

93. Wagner de Almeida, *Antropologia dos archivos*, 81.

94. These conflicts came to light above all at the Rio Conference in 1992, when various indigenous organizations drafted an alternative *Carta da Terra* in order to distinguish their views and claims from the official declaration.

95. Likewise, Radkau has pointed out the enduring "confused and contradictory picture" of the western wilderness conservation movement with respect to the struggles of indigenous people. Radkau, *Age of Ecology*, 380.

96. This is revealed, for example, in the stark contrasts drawn between indigenous people who are "close to nature" and those who are "assimilated," between "destructive" and "sustainable" development, between "valuable" (biodiverse) and "worthless" (degraded) nature.

97. Alvaro Tukano, "Uma reflexão sobre o movimento indígena do Brasil," 1993 (Private archive of the Rainforest Foundation, New York).

98. Slater, *Entangled Edens*, 148.

99. Revkin, *Burning Season*, 271.

100. Cf. Anthony B. Anderson and Ricardo Arnt, *O destino da floresta: Reservas extrativistas e desenvolvimento sustentável na Amazônia* (Rio de Janeiro: Relume Dumará, 1994).

101. The most well-known social science studies from those years include Stephen G. Bunker, *Underdeveloping the Amazon: Extraction, Unequal Exchange, and the Failure of the Modern State* (Urbana: University of Illinois Press, 1985); Catherine Caufield, *In the Rainforest* (New York: Knopf, 1985); Marianne Schmink and Charles Wood, eds., *Frontier Expansion in Amazonia* (Gainesville: University of Florida Press, 1984).

103. Cf. Michael Flitner, "Gibt es einen 'deutschen Tropenwald'? Anleitung zur Spurensuche," in *Der deutsche Tropenwald: Bilder, Mythen, Politik*, ed. Michael Flitner (Frankfurt am Main: Campus, 2000), 9–20.

104. It was not until the end of the 1980s that the rain forest topic entered the political arena of the Federal Republic of Germany.

105. U.S. Interagency Task Force on Tropical Forests, ed., *The World's Tropical Forests: A Policy, Strategy, and Program for the United States: Report to the President, US Policy and Strategy Program* (Washington, DC: Department of State, 1980). In the Brazilian case, the government established an interministerial working group on tropical forests in 1979.

106. German Bundestag, ed., *Protecting the Earth's Atmosphere: An International Challenge. Interim Report of the Study Commission of the 11th German Bundestag 'Preventive Measures to Protect the Earth's Atmosphere'* (Bonn: Deutscher Bundestag Referat Öffentlichkeitsarbeit, 1989).

107. German Bundestag, *Protecting the Tropical Forests*, 114.

108. Rainforest Alliance (1987, United States), Rainforest Action Network (1985, United States), World Rainforest Movement (1986, Malaysia), Rainforest Information Centre (1979, Australia), Rettet den Regenwald (1988, West Germany), Pro Regenwald (1989, West Germany), Oroverde (1989, West Germany), Arbeitsgemeinschaft Regenwald und Artenschutz (1987, West Germany).

109. Margaret Thatcher, "Speech Opening Saving the Ozone Layer Conference," London, March 5, 1989 (http://www.margaretthatcher.org/document/107593). Helmut Kohl, *Address to the International Environment Conference at The Hague*, The Hague, March 11, 1989 (Bonn: Deutscher Bundestag Referat Öffentlichkeitsarbeit, 1989).

110. "So isses," *Der Spiegel 48* (1988), 45.

111. John R. Mcneill, *Something New Under the Sun: An Environmental History of the Twentieth-Century World* (New York: Norton, 2000), 263.

112. Cf. José A. Pádua, "Nature and Territory in the Making of Brazil," in "New Environmental Histories of Latin America and the Caribbean," ed. José A. Pádua, Claudia Leal, and John Soluri, *RCC Perspectives* 7 (2013), 33–39.

113. Doug Boucher, Sarah Roquemore, and Estrellita Fitzhugh, "Brazil's Success in Reducing Deforestation," *Tropical Conservation Science* (Special Issue) 6, no. 3 (2013), 441.

114. Pádua, "Nature and Territory," 38. The geographer Matthew C. Hansen from the University of Maryland stated that "Brazil is a global exception in terms of forest change, with a dramatic policy-driven reduction in Amazon Basin deforestation." Matthew Hansen et al., "High-Resolution Global Maps of 21st-Century Forest Cover Change," *Science* 342, no. 6160 (November 15, 2013), 852.

115. Pádua, "Nature and Territory," 38.

116. Carlos A. Klink and Adriana G. Moreira, "Past and Current Human Occupation and Land Use," *The Cerrados of Brazil: Ecology and Natural History of a Neotropical Savanna*, ed. Paulo S. Oliveira and Robert J. Marquis (New York: Columbia University Press, 2002), 73–75.

117. Conference "Chico Vive: The Legacy of Chico Mendes and the Global Grassroots Environmental Movement," organized by Linda Rabben et al. http://www.cultur alsurvival.org/sites/default/files/programfinal031914.pdf.

118. Jonathan Watts, "Amazon Deforestation Report Is Major Setback for Brazil ahead of Climate Talks," *The Guardian*, November 27, 2015.

119. Anne Vigna, "São Paulo: eine Stadt auf dem Trockenen," *Le monde diplomatique*, April 9, 2015.

120. "O aumento no Desmatamento na Amazônia em 2013: um ponto fora da curva ou fora de controle?" http://imazon.org.br/publicacoes/o-aumento-no-desmat amento-na-amazo%CC%82nia-em-2013-um-ponto-fora-da-curva-ou-fora-de -controle.

121. "O Brasil atingirá sua meta de redução do desmatamento?" http://imazon.org .br/o-brasil-atingira-sua-meta-de-reducao-do-desmatamento/.

122. Cf. John R. McNeill and Peter Engelke, *The Great Acceleration: An Environmental History of the Anthropocene since 1945* (Cambridge, MA: Harvard University Press, 2016).

6. GREENPEACE AND THE BRENT SPAR CAMPAIGN

1. Daniel Yergin, *The Price: The Epic Quest for Oil, Money and Power*, rev. ed. (New York: Free Press, 2008), 651.

2. Cf. B.G.S. Taylor and R.G.H. Turnbull, "Development," in *North Sea Oil and the Environment*, ed. William J. Cairns, 57–77 (Barking: Elsevier, 1992), 76.

3. Tony Rice and Paula Owen, *Decommissioning the Brent Spar* (London: Spon, 1999), 15. See also Taylor and Turnbull, "Development."

4. Wolfgang Mantow, *Die Ereignisse um Brent Spar in Deutschland* (Hamburg: Deutsche Shell AG, 1995), 7.

5. Rice and Owen, *Decommissioning*, 16–21.

6. For a list of studies conducted by Shell Expro, see Mantow, *Die Ereignisse*, 13.

7. For more details on the list of disposal options, see Rice and Owen, *Decommissioning*, 30–41.

8. Rudall Blanchard Associates, "Brent Spar Abandonment BPEO," Commissioned for Shell U.K. Exploration and Production, December 1994, cited in Harald Berens, *Prozesse der Thematisierung in publizistischen Konflikten: Ereignismanagement, Medienresonanz und Mobilisierung der Öffentlichkeit am Beispiel von Castor und Brent Spar* (Wiesbaden: Westdeutscher Verlag, 2001), 115.

9. On the legal background of the incident, see Frank Biermann, "Weltumweltpolitik auf den sieben Meeren: Von der Meeresnutzungs- zur Meeresschutzordnung," in *Weltumweltpolitik*, ed. Udo E. Simonis, 197–216 (Berlin: Sigma, 1996).

10. Article 60 (3) makes it clear: "Any installations or structures which are abandoned or disused shall be removed to ensure safety of navigation. . . . Such removal shall also have due regard to fishing, the protection of the marine environment and the rights and duties of other States." UN Law of the Sea agreement of 1982 (http://www.un.org/depts/los/convention_agreements/texts/unclos/UNCLOS-TOC.htm; accessed October 20, 2015).

11. Convention on the Law of the Sea and IMO-regulations, cited in J. C. Side, "Decommissioning and Abandonment of Offshore Installations," in Cairns, *North Sea Oil*, 524.

12. Side, in Cairns, *North Sea Oil*, 529.

13. Cf. Mantow, *Die Ereignisse*, 12. Mantow also reproduces an article from the Greenpeace magazine. The statistics on the quantity of toxic substances was at first wildly underestimated: the first data suggested around 30 tons. The idea that sinking Brent Spar could set a precedent had already been mentioned: however, they initially assumed that only 60 rigs could follow suit—later they believed it could be as many as 400 other rigs, which could be disposed of in a similar way.

14. Frank Zelko, *Make It a Green Peace! The Rise of Countercultural Environmentalism* (New York: Oxford University Press, 2013).

15. See also Susanne Ramthun, "Aktionsraum Nordsee: Kampagnenpolitik an drei Fallbeispielen," in *Greenpeace auf dem Wahrnehmungsmarkt: Studien zur Kommunikationspolitik und Medienresonanz*, ed. Christian Krüger and Matthias Müller-Hennig, 117–33 (Münster: Lit, 2000).

16. Gerhard I. Timm, *Die wissenschaftliche Beratung der Umweltpolitik: Der Rat von Sachverständigen für Umweltfragen* (Wiesbaden: Deutscher Universitäts-Verlag, 1989), 298–99.

17. Michael Strübel, *Internationale Umweltpolitik: Entwicklungen-Defizite-Aufgaben* (Opladen: Leske und Budrich, 1992), 105–13.

18. Gijs Thieme, cited in Jochen Vorfelder, *Brent Spar oder die Zukunft der Meere: Ein Greenpeace-Report* (Munich: C. H. Beck 1995), 48.

19. On the affective nature of power see, for example, Sigrid Baringhorst, "'Sei ohne Sorge': Umweltschutz als Thema kommerzieller Werbung," in *Politische Inszenierung im 20. Jahrhundert*, ed. Sabine R. Arnold, Christian Fuhrmeister, and Dietmar Schiller, 160–70 (Vienna: Böhlau, 1998), 167.

20. Ulrich Jürgens, "Sieger sehen anders aus," in *Das Greenpeace Buch*, ed. Greenpeace, 281–89 (Munich: Beck, 1996), 283.

21. Christian Krüger and Matthias Müller-Hennig, "Wahrnehmungsprozesse, Kommunikationspolitik: Greenpeace als Unikum und Exempel," in Krüger and Müller-Hennig, *Greenpeace*, 9–15, here 9.

22. See Baringhorst, "'Sei ohne Sorge,'" 166.

23. Vorfelder, *Brent Spar*, 49.

24. Vorfelder, *Brent Spar*, 54.

25. Christopher Rootes, "Acting Globally, Thinking Locally? Prospects for a Global Environmental Movement," in *Environmental Movements: Local, National and Global*, ed. Christopher Rootes, 290–311 (Ilford: Cass, 1999), 300.

26. Vorfelder, *Brent Spar*, 56.

27. Vorfelder, *Brent Spar*, 87.

28. The *Spiegel* journalist Michaela Schießl recalls this situation in a TV documentary (28:54): *Duell auf hoher See: Der Kampf um die Brent Spar*, ARD, June 15, 2015 (http://www.ardmediathek.de/tv/Reportage-Dokumentation/Geschichte-im -Ersten-Duell-auf-hoher-Se/Das-Erste/Video?documentId=28994162&bcas tId=799280; accessed October 19, 2015).

29. Berens, *Prozesse*, 35–37.

30. Vorfelder, *Brent Spar*, 89.

31. Cf. Mantow, *Die Ereignisse*, 39.

32. On the discussions at the North Sea Conference, see Rice and Owen, *Decommissioning*, 86–88.

33. Patrick Scherler, *Management der Krisenkommunikation: Theorie und Praxis zum Fall Brent Spar* (Basel: Helbing und Lichtenhahn, 1996), 314.

34. Rice and Owen, *Decommissioning*, 4.

35. Klaus-Peter Johanssen, "'Wir kümmern uns mehr als um Autos': Die Geschichte einer Kampagne," in *PR-Kampagnen: Über die Inszenierung von Öffentlichkeit*, ed. Ulrike Röttger, 4th rev. ed., 315–21 (Wiesbaden: Springer, 2009), 317.

36. Poster, in Vorfelder, *Brent Spar*, 122.

37. Rice and Owen, *Decommissioning*, 89.

38. Press release from the German branch of Greenpeace, June 18, 1995, in Greenpeace, *Brent Spar und die Folgen, Analysen und Dokumente zur Verarbeitung eines gesellschaftlichen Konfliktes* (Göttingen: Die Werkstatt, 1997), doc. 52.

39. Press release from the DNV, October 18, 1995, in Greenpeace, *Brent Spar*, doc. 58.

40. During this examination, the DNV asked Greenpeace not only for the samples but also for three bins full of toxic refuse, which had allegedly been hidden away on Brent Spar in 1980 or 1981. Greenpeace UK made this accusation on June 16, after a former oil worker was in touch with the information that he had been involved in making this transaction. The informant's reports proved unverifiable. For more details on these mysterious toxic containers and the informant, see Rice and Owen, *Decommissioning*, 26–28.

41. Cf. Reiner Luyken, "Die Protestmaschine," *Die Zeit*, no. 37 (September 6, 1996), 16.

42. Melchett to Fay, September 4, 1995. Letter reprinted in Greenpeace, *Brent Spar*, doc. 50.

43. See also the commentary on the media response in bfp-analysis, "Brent Spar: Eine Falschmeldung und ihre Karriere. Über den Bau von Geschichtsbildern durch kognitive Ignoranz und kommunikative Penetranz," in Krüger and Müller-Hennig, *Greenpeace*, 205–22, here 207–10.

44. Greenpeace, *Brent Spar*, 36.

45. Reproduction of the advertisement in Greenpeace, *Brent Spar*, doc. 65.

46. Greenpeace, *Brent Spar*, doc. 65.

47. Henning von Vieregge, "'Sorry, wir haben einen Fehler gemacht!' Chancen und Risiken von Entschuldigungskampagnen," in Röttger, *PR-Kampagnen*, 327–31, here 328.

48. Results of the Emnid Survey in *Spiegel Special* (November 1995, 8, cited in Elisa-

beth Klaus "Die Brent-Spar Kampagne oder: Wie funktioniert Öffentlichkeit?" in Röttger, *PR-Kampagnen*, 97–119, here 102.

49. Thomas Schultz-Jagow, "Pyrrhus und Pyrrhus: eine kleine Bilanz der Mururoa-Kampagne," in Greenpeace, *Das Greenpeace Buch*, 37–43, here 39.

50. Greenpeace, *Brent Spar*, 26. The Mururoa campaign is also seen as a successful action that could have been more helpful than Brent Spar to Greenpeace's profile and issue positioning.

51. See also, for example, Dirk Kurbjuweit, "Die Robben sind uns näher," *Die Zeit* 47 (November 17, 1995).

52. Greenpeace, *Brent Spar*, 46.

53. *Deutsches Sonntagsblatt*, no. 29, June 23, 1995.

54. Ulrich Beck, "Weltrisikogesellschaft, Weltöffentlichkeit und globale Subpolitik: Ökologische Fragen im Bezugsrahmen fabrizierter Unsicherheiten," in *Umweltsoziologie* (Special Edition *Kölner Zeitschrift für Soziologie und Sozialpsychologie*, vol. 36), ed. Andreas Diekmann and Carlo Jaeger, 119–47 (Wiesbaden: Westdeutscher Verlag, 1996), 137, 139.

55. Conversation with Eric Hobsbawm, "Greenpeace, das ist die Revolution der Reichen," *Frankfurter Allgemeine Zeitung*, June 24, 1995, in Greenpeace, *Brent Spar*, doc. 9.

56. Hobsbawm, in Greenpeace, *Brent Spar*, doc. 9.

57. Nicola Liebert, "Wer rettet die Flußauen?" *die tageszeitung*, July 8 1995.

58. Cf. Bernhard Knappe, *Das Geheimnis von Greenpeace* (Vienna: Orac, 1993), 149–52

59. The sociologist Christopher Rootes maintains: "Effective action at national, let alone transnational, level necessitates organisation and a degree of elite autonomy. At anything approaching a global level, internally democratic social movement organisation is impractical given the need for speed of response and the obstacles to effective communication amongst members of a multinational constituency" (Rootes, "Acting Globally," 303).

60. On the changes made to the structures and strategies of NGOs internationally in the 1990s, see Ingo Take, *NGOs im Wandel: Von der Graswurzel auf das diplomatische Parkett* (Wiesbaden: Westdeutscher Verlag, 2002).

61. Franziska Sperfeld, "Strukturdaten großer Umweltverbände in Deutschland im Zeitraum 1990 bis 2010," *Unabhängiges Institut für Umweltfragen* 75, no. 1 (2014), 24–29.

62. "Angst vor der Endzeit," *Der Spiegel* 39 (September 25, 1995), 44.

63. See also Michael Hollmann, "Abrüstung auf dem Nordseegrund," *Frankfurter Rundschau*, January 22, 2003.

64. In the documentary by Wolfgang Luck, *Vom Ende einer Bohrinsel* (first broadcast on Phoenix on April 24, 2003 at 1:30 p.m.), the report mentions that the spectacular deconstruction of a drilling rig in a Kiel bay was the first technical attempt and was intended to serve as an ecological model for future disassemblies. However, marine biologists in the bay area maintained that the process had harmed the environment more than it had protected it. English marine biologists maintained that the drilling stations—and also the rest of Brent Spar—was the habitat for a type of coral that was on the red list of endangered species. Greenpeace replied correctly that it was not so much the coral itself but the reefs, the original habitat of the coral, that were threatened with destruction. It was also hard to argue that we must protect every car wreck that had moss growing all over it, as an important habitat. See also the article, "Leave Oil Rigs in the North Sea," *Guardian*, May 29, 2017.

65. Greenpeace clearly achieved this goal. In an analysis conducted by the journal *ÖKO-Test* on the transparency of income from donations, it got a good score, better than the German environmental organizations such as BUND and Nabu (*Öko-Test*, November 2002).

66. Greenpeace, ed., *Taten statt Warten: 25 Jahre Greenpeace Deutschland* (Hamburg, 2005).

67. Krüger and Müller-Hennig, *Greenpeace*, 205–22, see 207–8; Greenpeace, *Brent Spar*.

68. Dirk Maxeiner and Michael Miersch, *Öko-Optimismus* (Düsseldorf: Metropolitan, 1996), 39. It should be noted that the authors made some errors: they speak of a planned sinking in the *North Sea* (p. 36) and claim that Greenpeace organized its campaign around the figure of 5,000 tons (p. 37).

69. Ivar A. Aune and Nikolaus Graf Praschma, *Greenpeace: Umweltschutz ohne Gewähr,* (Melsungen: IAP—Institut zur Analyse der Protestindustrie, 1996), 95.

70. Burkhard Müller-Ullrich, *Medienmärchen: Gesinnungstäter im Journalismus* (Munich: Karl Blessing, 1996), 110.

71. Michael Kröher, "Wir lagen vor Mururoa," *die tageszeitung*, September 7, 2002.

72. On serious attacks by the economic and industrial lobby on the environmental movement, see Andrew Rowell, *Green Backlash: Global Subversion of the Environmental Movement* (London: Routledge 1996).

73. Rice and Owen, *Decommissioning*, 43–44.

74. Rice and Owen, *Decommissioning*, 143–45.

75. On the political decision-making process, see Grant Jordan, *Shell, Greenpeace and the Brent Spar* (Basingstoke: Palgrave 2001).

76. See also the Greenpeace ad of June 1, 1995, which portrayed an open can of sardines: "Thanks to Shell, all fish will soon be swimming in oil. The oil corporation plans to sink a polluted oil rig into the sea" (cited in Mantow, *Die Ereignisse,* 42).

77. Quote from Svenja Koch: "'Oma erzählt vom Krieg' and Polizistenlob: Heritage Communication bei Greenpeace," in *Tradition kommunizieren: Das Handbuch der Heritage Communication,* ed. Heike Bühler and Uta-Micaela Dürig, 163–70 (Frankfurt am Main: Frankfurter Allgemeine Buch, 2008), 166.

78. Greenpeace, ed., *Brent Spar und die Folgen: Zehn Jahre danach* (Hamburg, 2005).

79. Greenpeace, *Brent Spar und die Folgen,* 28.

80. Koch, "'Oma erzählt vom Krieg,'" 167.

81. See among others Meena Ahmed, *The Principles and Practice of Crisis Management: The Case of the Brent Spar* (Basingstoke: Palgrave Macmillan 2006). Daniela Puttena, *Praxishandbuch Krisenkommunikation: Von Ackermann bis Zumwinkel: PR-Störfälle und ihre Lektion* (Wiesbaden: Gabler Verlag, 2009). Ragnar Löftstedt and Ortwin Renn, "The Brent Spar Controversy: An Example of Risk Communication Gone Wrong," *Risk Analysis* 17, no. 2 (1996), 131–32.

82. For its own analysis of the failed communication strategy, see Shell International, Brent Dossier (http://www.shell.co.uk/sustainability/decommissioning/brent-spar-dossier.html; accessed October 18, 2015).

83. "20 Jahre Brent Spar," *Greenpeace Nachrichten,* no. 2 (2015), 8–11.

84. Greenpeace, ed., *20 Jahre nach Brent Spar: Offshore Öl- und Gasförderung im Nordostatlantik* (Hamburg, 2015).

7. A LANDSCAPE OF MULTIPLE EMERGENCIES

1. Officially Kashmir also denotes a larger area that includes the province of Jammu and Kashmir (subdivided into Jammu, Kashmir, and Ladakh divisions) administered by India, the Pakistan-administered territories of Azad Kashmir and Gilgit-Baltistan, and the unpopulated area of Aksai Chin at the western end that is controlled by China.

2. Mridu Rai contends that the British handed over Kashmir to the Dogras, as they wished "to build a buffer to prevent a uniting of Muslim interests, Kashmiri with Afghan," in the northwestern edge of India. *Hindu Rulers, Muslim Subjects: Islam, Rights and the History of Kashmir* (New Delhi: Permanent Black, 2004), 142.

3. See Shalini Panjabi, "From the Centre to the Margin: Tourism and Conflict in

Kashmir," in *Asia on Tour: Exploring the Rise of Asian Tourism*, ed. Tim Winter, Peggy Teo, and T. C. Chang, 223–38 (London: Routledge, 2009).

4. Ananya Jahanara Kabir privileges "desire" as a central element in the geopolitical claims over the territory. *Territory of Desire: Representing the Valley of Kashmir* (Minneapolis: University of Minnesota Press, 2009).

5. Michelle Maskiell, "Consuming Kashmir: Shawls and Empires, 1500–2000," *Journal of World History* 13, no. 1 (1992), 27–65.

6. The fascination of the Mughals with Kashmir was evident in the development of a category of Persian topographical poetry in the courts, specially extolling the natural beauty of the valley. (Chitralekha Zutshi, *Languages of Belonging: Islam, Regional Identity, and the Making of Kashmir* [New Delhi: Permanent Black, 2003], 29–34; Sunil Sharma, "Kashmir and the Mughal Fad of Persian Pastoral Poetry," in *Borders: Itineraries on the Edges of Iran*, ed. Stefano Pellò, 183–202 [Venice: Edizioni Ca Foscari Digital Publishing, 2016]).

7. Jahangir's apocryphal description of the valley as "paradise on earth" remains its most quoted depiction.

8. Attilio Petruccioli, "Gardens and Religious Topography in Kashmir," Environmental *Design: Journal of the Islamic Environmental Design Research Centre* 1–2 (1991), 64–73.

9. *Lalla Rookh (1817) by Thomas Moore (1779–1852)* (http://www.columbia.edu/itc/mealac/pritchett/00generallinks/lallarookh/).

10. Brigid Keenan, *Travels in Kashmir: A Popular History of Its People, Places, and Crafts* (Delhi: Oxford University Press, 2006), 93–95; Ashok Pratap Malhotra, *Making British Indian Fictions: 1772–1823* (New York: Palgrave Macmillan, 2012), 73–77.

11. Sandeep Banerjee, "'Not Altogether Unpicturesque': Samuel Bourne and the Landscaping of the Victorian Himalaya," *Victorian Literature and Culture* 42 (2014), 351–68.

12. The idea of houseboats was not introduced by the British as is claimed. Small houseboats called *doongas* were being used by poor families to live on the Jhelum and on the Dal. Their development into ornate, luxurious houseboats, was the result of "the interaction between the houseboat owner, his European boarder and the local craftsman" (interview with Saleem Beg, November 9, 2006).

13. Mridu Rai, "Making a Part Inalienable: Folding Kashmir into India's Imagination," in *Until My Freedom Has Come: The New Intifada in Kashmir*, ed. Sanjay Kak, 250–78 (New Delhi: Penguin Books, 2011).

14. Kabir also marks the lake's visual centrality with reference to the Hindi film *Roja*, set amid the violence of the 1990s in Kashmir. The conditions did not permit shooting in the valley though, so the film was shot in the adjoining mountainous provinces with similar vistas. However, as Kabir states, "the absence of the signature lake and shikara reveal[ed] a central lack that disfigure[d] the entire narrative" (Kabir, *Territory of Desire*, 45).

15. Walter R. Lawrence, *The Valley of Kashmir* (Srinagar: Gulshan Books, 2005 [originally published London: Henry Frowde, 1895], 345.

16. Amy Waldman, "Border Tension a Growth Industry for Kashmir," *New York Times*, October 18, 2002.

17. Tasaduq H. Shah et. al., "Elevating Fishers of Dal Lake in Jammu and Kashmir," *Journal of the Indian Fisheries Association* 30 (2003), 173–82.

18. Lawrence, *Valley of Kashmir*, 22.

19. These include the Srinagar Master Plan of 1971, Lake Area Master Plan by Stein (1972), Enex Consortium Report (1978), and the Dal Lake Development Report by Riddle (1985). Later proposals have included the Dal Lake Conservation Plan by Iram Consultants (1997) and a report prepared by the Alternate Hydroelectric Department of the University of Roorkee (2000).

20. LAWDA is charged with overseeing, managing, and conserving the water bodies and waterways of the state of Jammu and Kashmir. Its current focus is on the management of the Dal/Nigeen Lakes, using funds from the National Lake Conservation Plan, the Prime Minister's Reconstruction Programme.

21. Ishfaq Tantry, "Admit Dal Area Has Shrunk: High Court to Lake Authority," *Greater Kashmir*, May 23, 2017.

22. "Dal Needs 'Political Will' [editorial]," *Greater Kashmir*, May 16, 2007.

23. Shahab Fazal and Arshad Amin, "Hanjis Activities and Its Impact on Dal Lake and Its Environs," *Research Journal of Environmental and Earth Sciences* 4, no. 5 (2012), 511–24; MRD Kundangar, "Squeezing Dal Lake: Historical Perspective," *Greater Kashmir*, June 5, 2012; Manzoor Ahmad Khan, "Dal Lake of Kashmir: Problems, Prospects and Perspectives," *International Journal of Multidisciplinary Research and Development* 2, no. 2 (2015), 462–69.

24. Arif Shafi Wani, "Government Seeks Public Response on Actual Size of Dal Lake," *Greater Kashmir*, December 28, 2015.

25. "Land mass" included the area under cultivation and habitation.

26. LAWDA credits this to its own efforts to remove "habitations, land mass and vegetation in the water body" (Wani, "Government Seeks Public Response").

27. Fazal and Amin, "Hanjis Activities," 515–16. For the study, "water body" was taken

to mean the water area of the lake excluding the marshy area and floating garden; while "marshy area" referred to areas under the cover of weeds, where water-related activities like the extraction of aquatic foods was being undertaken.

28. Fazal and Amin, "Hanjis Activities"; M. H. Wani et al., "Economic Valuation and Sustainability of Dal Lake Ecosystem in Jammu and Kashmir," in *Knowledge Systems of Societies for Adaptation and Mitigation of Impacts of Climate Change*, ed. Sunil Nautiyal et al., 95–118 (Berlin: Springer-Verlag, 2013).

29. Maroosha Muzaffar, "Dal Lake = Srinagar's Sewage Dump?" *Indian Express*. April 3, 2009.

30. Personal interviews with Haji Ghulam Rasool, Mohammed Tuman, and other Hanjis.

31. Mona Bhan and Nishita Trisal, "Fluid Landscapes: The Politics of Conservation and Dislocation," *Greater Kashmir*, June 30, 2009.

32. Bhan and Trisal, "Fluid Landscapes."

33. A. Khan, "Valley's Floating Marvels Dying a Slow Death," *Tribune*, June 9, 2016.

34. There are many instances of such conflicts in India that reveal fractures over notions of "conservation" and "development." The most pertinent example here is the protest against the eviction and clearing of the Loktak Lake in northeast India, which also has habitations on floating islands and is the refuge of the endangered Manipur brow-antlered deer. See Monica Amador, "Caught between War, Development and Conservation: The Case of Loktak Lake in Northeast India," *Intercultural Resources* (2011) (https://www.ritimo.org/Caught-between -War-Development-and-Conservation-the-Case-of-Loktak-Lake-in).

35. Gunnel Cederlof and Kalyanakrishnan Sivaramakrishnan, "Ecological National-isms: Claiming Nature for Making History," in *Ecological Nationalisms: Nature, Livelihoods and Identities in South Asia*, ed. Cederlof and Sivaramakrishnan, 1–40 (New Delhi: Permanent Black, 2005).

36. Arif Shafi Wani, "LAWDA's Prescription for Clean Dal: Keep Water Level High," *Greater Kashmir*, April 12, 2007.

37. Wani et al., "Economic Valuation," 109.

38. Neha Wajahat Qureshi and M. Krishnan, "Lake Fisheries in Kashmir: A Case More Undone than Done," *Economic and Political Weekly* 1, no. 2 (2015), 66–69, here 67.

39. A. F. Robertson, "The Dal Lake: Reflections on an Anthropological Consultancy in Kashmir," *Anthropology Today* 3, no. 2 (1987), 7–13, here 7.

40. The Wildlife (Protection) Act, 1972, p. 21 (http://lawmin.nic.in/ld/P-ACT/1972/ The%20Wild%20Life%20(Protection)%20Act,%201972.pdf).

41. The symbolism of the hangul resonates in different ways with Kashmiris, as in the work of Malik Sajad, *Munnu: A Boy from Kashmir* (London: Fourth Estate, 2015). In a graphic novel about growing up in Kashmir during the violence, he represents Kashmiris as humanoids of the hangul—the precarious existence of the hangul serving as a metaphor for the plight of the besieged Kashmiris.

42. Ashiq Hussain, "Endangered Hangul's Population Decreases by 40% in Kashmir," *Hindustan Times*, August 24, 2016.

43. Farooq Shah, "Army Violates National Park Rules 'in Pursuit of Burhan,'" *Kashmir Observer*, May 6, 2016.

44. Burhan Wani, the son of a college principal from a village in south Kashmir had emerged as the poster boy for the new wave of militancy. His many posts and videos on social media—showing him as equally comfortable playing cricket or holding automatic weapons—made him hugely popular in the valley.

45. Ayesha Siddiqi, "Kashmir and the Politics of Water," *Al Jazeera English*, August 1, 2011.

46. India began building major hydropower projects in Kashmir in the 1970s and now has thirty-three projects at various stages of completion on the rivers in Kashmir.

47. Baba Umar, "Kashmir: A Water War in the Making?" *Diplomat*, June 9, 2016.

48. Mukeet Akmali, "Dal Vegetables Suffice Srinagar's Demand," *Greater Kashmir*, September 20, 2016; Abhishek Saha, "Through Thick and Thin: Market on Dal Lake Feeds Curfew-hit Srinagar," *Hindustan Times*, August 26, 2016.

49. Mudasir Yaqoob, "Weeds, Red Algal Bloom Deface Dal," *Greater Kashmir*, August 30, 2016; Riyaz Wani, "Fresh Algae Blooms Threaten the Dal Lake," Tehelka.com, September 20, 2016 (http://www.tehelka.com/2016/09/fresh-algae-blooms-threaten-the-dal-lake/).

50. Thomas Greider and Lorraine Garkovich, "Landscapes: The Social Construction of Nature and the Environment," *Rural Sociology* 59, no. 1 (1994), 1–24.

8. THE ADIVASI VERSUS COCA-COLA

1. "University of Michigan bans Coca-Cola sales," *Associated Press*, December 30, 2005.

2. Another important cause of the protests was that Coca-Cola was suspected of having worked with paramilitaries in Colombia, who had murdered trade union representatives.

3. Cf. the company's 2005 consolidated balance sheet (https://www.sec.gov/Arch ives/edgar/data/21344/000104746906002588/a2167326z10-k.htm; accessed January 24, 2018).

4. The Indian historian Vinay Lal provides further background on this in "Coca Cola in India," AsiaMedia, September 1, 2006 (http://www.sscnet.ucla.edu/south asia/History/Current_Affairs/coca_cola_india.html; accessed January 24, 2018). In the company's own account, this phase is largely overlooked. See the home-page of Coca-Cola India (www.coca-colaindia.com). Coca-Cola later alleged that it withdrew from India because, had it stayed, it would have had to reveal its secret recipe to the Indian government. By contrast, the fact that ownership of the company had to be 40 percent Indian often went unmentioned. Mark Pendergast follows the company's official account on this point in *For God, Country and Coca Cola: The Unauthorized History of the Great American Soft Drink and the Company that Makes It* (New York: Scribner, 1993), 317, 398. Constance L. Hays talks of a difficult reentry into the Indian market, in *The Real Thing: Truth and Power at the Coca-Cola Company* (New York: Random House Trade Paperbacks, 2004), 323, 331). For a critical account with background on the withdrawal, see C. Gopinath and Anshuman Prasad, "Towards a Critical Framework for Understanding MNE Operations: Revisiting Coca-Cola's Exit from India," *Organization* 20, no. 2 (2013), 212–32.

5. According to its own statements, the Coca-Cola Company invested more than $1 billion in India up to 2003 (it is less forthcoming about the profits that were transferred back). At this point it operated twenty-four of its own bottling plants and was involved via franchise in another twenty-five plants. Coca-Cola Company, "Coke Facts" (since deleted), but the archive is accessible (http://web.archive. org/web/20080705102422/http:/www.cokefacts.com/India/bg_in_history.shtml; accessed January 24, 2018).

6. This produced total earnings of around 1 million rupees (around €22,300 at the April 2002 exchange rate), a not inconsiderable increase in income for the Panchayat.

7. Gavin Rader, "The People of Plachimada vs. Coca-Cola and the Fight For Water Democracies in India," *Treganza Museum Anthropology Papers*, nos. 24–25 (2007–8), 148–52, here 149.

8. Coca-Cola Company, "Coke Facts" (https://web.archive.org/web/200807051023 58/www.cokefacts.com/India/facts_in_qa.shtml; accessed January 24, 2018). For an NGO statement, see Rohan D. Mathews, "The Plachimada Struggle against

Coca Cola in Southern India," July 1, 2011 (http://www.ritimo.org/article884.
html; accessed January 24, 2018).

9. Quoted in Rader, "People of Plachimada," 150.

10. Ajaykumar P. Panicker, "Counter-Hegemonic Collective Action and the Politics
of Civil Society: The Case of a Social Movement in Kerala, India, in the Context
of Neoliberal Globalization" (PhD diss., University of Miami, 2008), 115. Given
the extreme poverty in the Adivasi community, any loss of income had a severe
effect on their ability to feed their families.

11. Paul Brown, "Coca-Cola Plant Must Stop Draining," *Guardian*, December 19,
2003. The scarcity of water ultimately even affected cola production, causing
tankards to be driven to private wells and to buying the necessary water from
farmers. Panicker, "Counter-Hegemonic Collective Action," 125, 140–41. Coca-
Cola chose this region because water had been comparatively plentiful there. It
is unlikely that such a steep drop in the water table was caused by climatic influ-
ences—especially when we compare the area to neighboring districts.

12. For more on social conditions in Plachimada, see Panicker, "Counter-Hegemonic
Collective Action," 111–12. The "Adivasi" from Plachimada belonged to the Eravalar and Malasar ethnic groups. Other studies on Plachimada overlook the fact
that residents were very disappointed from the beginning, because the few jobs
went to party members and trade unionists who came from elsewhere, and were
only given to locals in exceptional cases (173–75). Local politicians also attributed
the first protests to this disappointment.

13. Sargam Metals Laboratories in Chennai. This laboratory was sold in 2013 to the
Bureau Veritas Consumer Product Services Division (BVCPS) (http://www
.prnewswire.com/news-releases/bureau-veritas-announces-sargam-laboratory-pri
vate-limited-name-change-215946051.html; accessed January 24, 2018). The health
authority also had the water tested by a state laboratory and instructed the Panchayat
on May 13, 2003, to inform the public that the water was undrinkable. C. R. Bijoy,
"Kerala's Plachimada Struggle," *Economic and Political Weekly* 14 (2006), 4332–39.

14. In 2001 E. K. Janu had been successful in a campaign to get the government of
Kerala to officially recognize the right of the Adivasi to their own landholdings
for the first time. She was therefore one of the most prominent Adivasi activists.
Cf. Mohamed Nazeer, "The Power Women: Seeking Justice in God's Own Land,"
The Hindu, September 3, 2011.

15. Local elites saw the actions of the Adivasi as a threat to the status quo, and took
partly violent action against their early protests. Panicker, "Counter-Hegemonic
Collective Action," 178–79.

16. Cf. P. Baburaj and C. Sarathchandran, *Thousand Days and a Dream* (Film, India, 2006), Bijoy, "Kerala's Plachimada Struggle"; Mathews, "Plachimada Struggle." Coca-Cola's attempt to get a legal ban on the protest camp opposite the factory entrance failed, but it did force the protest movement to invest a considerable proportion of its limited resources in paying for the legal battle—something that was no problem for Coca-Cola.

17. Panicker, "Counter-Hegemonic Collective Action," 123. For financial reasons, the workers had no choice but to sign the document.

18. India Resource Center, "CorpWatch India Responds to Coca-Cola" (http://www .indiaresource.org/campaigns/coke/2003/corpwatchindiaresponds.htm; accessed January 24, 2018).

19. Because the different ethnic groups that are labeled "Adivasi" make up just 1.5 percent of the population in the state of Kerala, they are virtually insignificant as a voting bloc.

20. Cf. Panicker, "Counter-Hegemonic Collective Action," 131. It should also not be overlooked that the drop in the water table meant that influential and prosperous farmers, who employed the farm laborers, now also had to contend with a loss of income. This fact, as well as a change in policy at party headquarters, must have been at least as important as the protesters' symbolic cleaning in bringing about the Panchayat's change of heart.

21. BBC Radio 4, "*Face the Facts* Investigates Coca-Cola Plant in India," July 24, 2003 (Radio broadcast July 25, 2003) (http://www.bbc.co.uk/pressoffice/pressreleases/ stories/2003/07_july/24/face_facts.shtml; accessed January 24, 2018). A London toxicologist and professor, John Henry, feared there would be a serious effect on people's health: "The results have devastating consequences for those living near the areas where this waste has been dumped and for the thousands who depend on crops produced in these fields."

22. BBC Radio 4, "*Fact the Facts* Investigates."

23. Coca-Cola Company, "The Coca-Cola Company Addresses Allegations Made about Our Business in India" (https://web.archive.org/web/20061007210957/ http://www2.coca-cola.com/presscenter/viewpoints_india_situation.html; accessed January 24, 2018).

24. Bijoy, "Kerala's Plachimada Struggle," 4335.

25. Drinking water ordinance from 2011; sewage sludge ordinance from 1992; Indian Standard Specification for Drinking Water, IS 10500 (1992) (http://hppcb.gov.in/ eiasorang/spec.pdf; accessed May 24, 2013). It should be pointed out that cad-

mium could theoretically find its way into drinking water by way of artificial fertilizers, an argument that Coca-Cola used on another occasion to explain the contamination of the water. This excuse could not be used with regard to the dried sludge.

26. P. N. Venugopal, "Coca Cola Moving out of Plachimada?" *India Together*, January 27, 2006.

27. Baburaj and Sarathchandran, *Thousand Days*, 00:36:32–00:36:57.

28. On the frequency of corruption in India, see Jennifer Bussell, *Corruption and Reform in India: Public Services in the Digital Age* (Cambridge: Cambridge University Press, 2012); Mira Fels, *Making Sense of Corruption in India: An Investigation into the Logic of Bribery* (Berlin: LIT, 2008).

29. Bijoy, "Kerala's Plachimada Struggle," 4335.

30. Coca-Cola, Allegations.

31. Centre for Science and Environment (CSE), "Analysis of Pesticide Residues in Soft Drinks," August 5, 2003 (http://www.cseindia.org/userfiles/SOFTDRINK. pdf; accessed January 24, 2018), 13. The CSE study also pointed to a lack of legal regulations, which meant that these worryingly high quantities were not illegal. The study did not so much attack the soda manufacturer as appeal to the government finally to introduce appropriate standards and to monitor them properly. The previous year, the institute had shown similar pesticide residues in Indian drinking water and mineral water; the soft-drinks study was not a campaign directed against the Coca-Cola Company, but a follow-up study to illustrate that the government needed to take action. In many other countries, the Coca-Cola Company faced a very different problem when it came to purification: how could the reusable bottles be cleaned in such a way that they met the relevant hygiene standards and could be refilled?

32. Jonathan Hills and Richard Welford, "Case Study: Coca-Cola and Water in India," *Corporate Social Responsibility and Environmental Management* 12 (2005), 168–77, here 170.

33. CSE, "PepsiCo and Coca Cola Conjure Up 'Data' that Seeks to Convolute, Confuse and Take the Indian Public for a Ride," August 7, 2003 (http://cseindia.org/node/508; accessed January 24, 2018).

34. CSE, "Analysis of Pesticide Residues in Soft Drinks," August 2, 2006 (http://www .indiaenvironmentportal.org.in/files/labreport2006.pdf; accessed January 16, 2018).

35. John Vidal, "Things Grow Better with Coke," *The Guardian*, November 1, 2004.

36. "Water Is Not a Private Property, Says Plachimada Declaration," *Hindu*, April 24, 2004.

37. R. Krishna Kumar, "Resistance in Kerala," *Frontline* 21, no. 3 (2004). The Indian and international press did not report to the same extent the fact that PepsiCo exhibited exactly the same sort of behavior in its dealings with the environment. However, it displayed exactly the same pattern of behavior: overuse of ground water, cadmium in sludge that was distributed to farmers, legal appeals against the decisions of the Panchayat, counter-reports by experts, and corruption of environmental officials (a remarkable similarity to the Coca-Cola case). Cf. M. Suchitra, "Inaction on Panel Findings against Beverage Major," *India Together*, August 6, 2008.

38. Cf. Paul Brown, "Coca-Cola in India Accused of Leaving Farms Parched and Land Poisoned," *Guardian*, July 25, 2003; Marc Pitzke, "Wut auf Brause-Multi: Umweltschützer geisseln Coca-Colas Wasserverbrauch," *Spiegel-online*, March 19, 2007 (http://www.spiegel.de/wirtschaft/wut-auf-brause-multi-umweltschuet-zer-geisseln-coca-colas-wasserverbrauch-a-472470.html; accessed January 24, 2018); Laurence Caramel, "A Bombay, Coca-Cola est érigée en symbole de ces mulinationales accusées de pollution," *Le Monde*, January 22, 2004; Vandana Shiva, "Coca Cola löscht den Durst nicht," *Le Monde Diplomatique*, March 11, 2005.

39. It would be useful to study the reception history either of this case or a similar one, and to initiate a piece of historical agenda research.

40. Hills and Welford, "Case Study," 171. Coca-Cola's image crisis was further exacerbated after the former Miss Universe and filmstar Sushimata Sen accepted $315,000 in compensation to drop a sexual harassment complaint against Coca-Cola's head of marketing (*Times of India*, December 9, 2003). The *Economic Times*, which always covered Coca-Cola in a very positive light, reported that sales of Coke had dropped by 40 percent in the supermarket chain Foodworld, while distributors from southern India reported a drop of up to 80 percent. "Cola Sales Enter Drop Zone, Lose 40% Altitude," *Economic Times*, August 13, 2003. This appears to contradict the figures in Coca-Cola's annual report (Coca-Cola Company, 2003 Summary Annual report [http://www.coca-colacompany .com/content/dam/journey/us/en/private/fileassets/pdf/unknown/unknown/ annual_2003.pdf; accessed January 16, 2018], 15), according to which the company grew by 18 percent in India between 2002 and 2003. It was not until the 2005 report (Coca-Cola Company, 2005 Annual Review [http://www.coca-co lacompany.com/content/dam/journey/us/en/private/fileassets/pdf/unknown/

unknown/koar_05_complete.pdf; accessed January 16, 2018], 26) that reference was made to sinking sales figures in India.

41. Hills and Welford, "Case Study," 171.

42. "Kirpal's Concern about Court Order on Coke," *The Hindu*, March 11, 2004 (http://www.thehindu.com/2004/03/11/stories/2004031112830500.htm; accessed January 24, 2018).

43. Sreedevi Jacob, "In the Dock" April 30, 2004 (http://www.downtoearth.org.in/news/in-the-dock-11161; accessed January 24, 2018).

44. The description of events in this section draws on C. R. Bijoy, "Kerala's Plachimada Struggle," 4335–37; Hills and Welford, "Case Study," 170–72; Panicker, "Counter-Hegemonic Collective Action," 147–53; Mark Thomas, *Belching Out the Devil: Global Adventures with Coca-Cola* (New York: Nation Books, 2008), 189–96.

45. Interview with Krishnan, the chairman of the Panchayat at the time, in Baburaj and Sarathchandran, *Thousand Days*, 00:26:25–00:27:23.

46. Kerala Panchayat Raj Act 1994 (http://sanitation.kerala.gov.in/wp-content/uploads/2017/07/the-kerala-panchayat-raj-act-1994.pdf; accessed January 16, 2018), 218. The Panchayat had the authority "to manage and regulate the minor irrigation, water management and water development."

47. A fierce legal critique of this judgment appears in Nishita Vasan, "The Plachimada Problem," November 13, 2009 (https://works.bepress.com/nishita_vasan/1/download; accessed January 24, 2018).

48. The controversy over the interpretation of the experts' report is reflected in the following publications: K. Ravi Raman, "Breaking New Ground: Adivasi Land Struggle in Kerala," *Economic and Political Weekly*, March 9, 2002, 916–19; Saleem Romani, "Plachimada Water," *Economic and Political Weekly*, December 3, 2005, 5134, 5211.

49. The problem was that Coca-Cola did not make any effort to address residents' complaints—not even about direct pollution by the plant. Three years and a day after the beginning of the protest (April 23, 2005), the Indian weekly magazine *Outlook* took a sample from a nearby well and had it examined by an independent institute, which also happened to do work for Coca-Cola. The water sample showed a PH-level of 3.5, and the quantity of total dissolved solids (TDS) was not at the permitted level of 2,000 but rather at 9,634. S. Anand, "Don't Poison My Well," *Outlook News Magazine*, May 16, 2005. A corresponding counter-report followed soon afterward: V. R. L. Murthy, "Allegations". At the time, Murthy was chief hydrologist for the Coca-Cola Company in India.

50. "Coca-Cola Seeks Nod for Shifting Plant Locale (from Plachimada in Perumatty Panchayat to an Industrial Centre in the Pundusseri Panchayat)," *India Business Insight*, January 3, 2006.

51. The CPI (Marxist) is the biggest of several leftist parties in India. It draws on Marxism, unlike the more radical CPI (Maoist) or the Naxalites. The CPI (Marxist) does not seek to dismantle the Indian state; instead, it takes part in elections as a democratic party, and participates in government coalitions, for example, to promote agrarian reform and to improve social conditions.

52. See the CPI (Marxist) statement on this topic (http://www.cpim.org; accessed October 19, 2015).

53. Cf. Betsy McKay, "Why Coke Aims to Slake Global Thirst for Safe Water," *Wall Street Journal*, March 13, 2007; Steve Stecklow, "How a Global Web of Activists Gives Coke Problems in India," *Wall Street Journal*, June 7, 2005. In this report, all the allegations against Coca-Cola are labeled as unproven.

54. Stecklow, "How a Global Web."

55. For more information on the conflict in Mehdiganj and Kaladera, see Thomas, *Belching With the Devil*, 197–220.

56. The Energy and Resource Institute (TERI), "Executive Summary on Independent Third Party Assessment of Coca-Cola Facilities in India," 13–14 (http://www.teriin.org/upfiles/projects/Coca-cola-ES.pdf; accessed January 24, 2018).

57. List of TERI's sponsors (http://www.teriin.org/index.php?option=com_content&task=view&id=43#; accessed May 23, 2013). Particularly in 2006 and 2007, Coca-Cola helped to fund some of the institute's educational projects, which, ironically, were focused on "Mobilising Youth for Water Conservation."

58. TERI, "Executive Summary Report," 21.

59. Coca-Cola India, "Towards Sustainability: Environment Report 2008–2009," (http://assets.coca-colacompany.com/70/f8/f41bf2064de5885a0d44e90165da/2008-2009_india.pdf; accessed May 24, 2013). Cf. Ashok Sharma, "Coca-Cola in India Pesticide Free," *USA Today*, January 15, 2008.

60. Quoted in Arneel Karnani, "Corporate Social Responsibility Does Not Avert the Tragedy of the Commons—Case Study: Coca-Cola India," Ross School of Business Papers 1173 (2012), 13.

61. Karnani, "Corporate Social Responsibility," 17–18, 26.

62. Most recently on this topic: Angelika Epple, "Lokalität und die Dimensionen des Globalen: Eine Frage der Relationen," *Historische Anthropologie* 21, no. 1 (2013), 4–25.

63. UN Committee on Economic, Social and Cultural Rights (CESCR), General Comment No. 15: The Right to Water (Arts. 11 and 12 of the Covenant), January 20, 2003, E/C.12/2002/11.

64. Chilmayi Shalya, "Plachimada Villagers Still Await President's Nod", *Times of India*, September 9, 2012. It is clear that Coca-Cola India fears that this would open the door to other cases, and if the tribunal were set up the company would certainly oppose it.

65. Local members of the movement expressed themselves to this effect in interviews. Cf. Baburaj and Sarathchandran, *Thousand Days*; Panicker, "Counter-Hegemonic Collective Action," 181–83.

66. Cf. Ananthakrishnan Aiyer, "The Allure of the Transnational: Notes on Some Aspects of the Political Economy of Water in India," *Cultural Anthropology* 22, no. 4 (2007), 640–58, here 649–52. Aiyer criticizes research into and representations of the Plachimada conflict because the issues of land reform and rural poverty, which are fundamental to the Adivasi, are pushed into the background.

67. Panicker, "Counter-Hegemonic Collective Action," 203–4.

68. On the significance of this diversity in the new global movement, see David Featherstone, *Resistance, Space and Political Identities: The Making of Counter Global Networks* (Oxford: Wiley-Blackwell, 2008), 132.

69. One could interpret Coca-Cola's strategy of constantly demanding new scientific studies for the authorities or for the courts as an attempt to "bleed" the movement financially.

70. Panicker, "Counter-Hegemonic Collective Action," 202–3.

71. "We believe in preserving and protecting water resources. Protecting and improving access to and the availability of water remains one of our long-term goals. We partner with many organizations, governments and local communities to develop and implement sustainable water initiatives around the world." Coca-Cola Company, 2005 Summary Annual Report, 3.

72. India Resource Center, "Coca Cola's Latest Scam: Water Neutrality," November 25, 2008 (http://www.indiaresource.org/campaigns/coke/2008/neutrality.html; accessed January 24, 2018). Cf. Fred Pearce, "Greenwash: Are Coke's Green Claims the Real Thing?" *Guardian*, December 4, 2008.

73. Venugopal, "Coca Cola Moving."

74. Gayatri Chakravorty Spivak, "Can the subaltern speak?" in *Marxism and the Interpretation of Culture*, ed. C. Nelson and L. Grossberg, 271–313 (Basingstoke: Macmillan Education, 1988).

CONTRIBUTORS

KENNETH ANDERS is an independent scholar and cofounder of the Büro für Landschaftskommunikation in Bad Freienwalde near Berlin.

BERND-STEFAN GREWE studied history, French, and philosophy at the universities of Trier and Paris X—Nanterre (PhD, University of Trier). He is chair and director of the Institute of Didactics of History and Public History at Tübingen University. Other research interests include environmental and rural history, history of colonialism and of global commodities and material culture with a special focus on South Africa, India, and Great Britain. He is the author of *Der versperrte Wald. Ressourcenmangel in der bayerischen Rheinpfalz 1815–1870* (Boehlau, 2004); "The London Gold Market 1910–1935," in Christof Dejung and Niels P. Peterson, eds., *Power, Institutions and Global Markets: Actors, Mechanisms and Foundations of World-Wide Economic Integration, 1850–1930,* (Cambridge University Press, 2013), 112–32; with Thomas Lange, *Kolonialismus: Quellen und Darstellungen* (Reclam, 2015), and with Karin Hofmeester, *Luxury in Global Perspective: Objects and Practices* (Cambridge University Press, 2016).

PATRICK KUPPER is a full professor of history and head of the Economic and Social History Unit at the University of Innsbruck, Austria. His expertise is in transnational economic, social, and environmental history of modern Europe. He is a board member of the International Association for Alpine History (IAAH) and of the Arcadia project featuring *Online Explorations in Environmental History* (http://www.environmentandso ciety.org/arcadia). His book publications include *Creating Wilderness: A Transnational History of the Swiss National Park* (Berghahn, 2014), with David Gugerli and Daniel Speich, *Transforming the Future: ETH Zurich and the Construction of Modern Switzerland 1855–2005* (Chronos, 2010), and with Bernhard Gissibl and Sabine Höhler, *Civilizing Nature: National Parks in Global Historical Perspective* (Berghahn, 2012).

KEVIN NIEBAUER studied history, Spanish and Latin American studies at the Free University of Berlin. He is a doctoral student at the International Research Training Group "Between Spaces: Movements, Actors and Representations of Globalisation" at the Institute for Latin American Studies in Berlin. His dissertation deals with the Brazilian Amazon rainforest as a transatlantic trope of ecological crisis.

SHALINI PANJABI is an independent researcher based in Bangalore, India. She completed her PhD in sociology at the Delhi School of Economics on aspects of private education in India, and has received a Postdoctoral Research Award from the Charles Wallace Trust. Her research interests are in the larger areas of orality, literacy, and cultural and natural heritage. One of her field sites has been Kashmir, where she has focused on complex issues concerning heritage, conservation, and development in a conflict situation. She is also currently involved in a project on Sindhi oral narratives in the border regions of western India.

ELKE SEEFRIED studied economics, history, and political science at Augsburg's University of Applied Sciences, Augsburg University and Friedrich-Alexander University at Erlangen-Nuremberg. She received a Diploma from the University of Applied Sciences in 1999 and a PhD from Augsburg University in 2004, where she continued to work for some years. She received fellowships from the German Historical Institutes London and Paris as well as from Historisches Kolleg Munich and received her Habilitation from Munich's Ludwig Maximilian University in 2013. Since 2014, she has been a professor of modern history at Augsburg University and second deputy director at the Institut für Zeitgeschichte Munich-Berlin

(Institute for Contemporary History). Her publications include *Zukünfte: Aufstieg und Krise der Zukunftsforschung* (2015, English edition with Berghahn Books in 2019), and "Rethinking Progress: On the Origin of the Modern Sustainability Discourse, 1970–2000," *Journal of Modern European History* 3 (2015).

FRANK UEKÖTTER studied history, political science, and the social sciences at the universities of Freiburg and Bielefeld in Germany, Johns Hopkins University in Baltimore, and Carnegie Mellon University in Pittsburgh. He received a PhD from Bielefeld University in 2001, where he continued to work for several years. He moved to Munich in 2006 and joined the Research Institute of the Deutsches Museum, taught at Munich's Ludwig Maximilian University, and helped to build the Rachel Carson Center for Environment and Society. Since 2013, he has been a reader in environmental humanities at the University of Birmingham (UK). His publications include *The Age of Smoke: Environmental Policy in Germany and the United States, 1880–1970* (University of Pittsburgh Press, 2009), *The Greenest Nation? A New History of German Environmentalism* (MIT Press, 2014), and, as editor, *The Turning Points of Environmental History* (University of Pittsburgh Press, 2010).

ANNA-KATHARINA WÖBSE is an environmental historian and research associate at the University of Gießen. She has extensively published on animal–human relations, visual history, the history of transnational environmental movements, and environmental diplomacy in the League of Nations and the United Nations. She is currently involved in a multidisciplinary research project exploring the transboundary history of European wetlands.

INDEX